CITIZENS BAND RADIO DIGEST

by william j. bradley

DBI BOOKS INC. NORTHFIELD, ILLINOIS

Citizens Band Radio Digest Staff

EDITOR:
William J. Bradley
CONSULTING EDITOR:
Joseph Schroeder
ART DIRECTOR:
Mary E. MacDonald
PRODUCTION MANAGER:
Pamela J. Johnson
ASSOCIATE PUBLISHER:
Sheldon L. Factor

Copyright © MCMLXXVI by DBI Books, Inc., 540 Frontage Rd., Northfield, Ill. 60093, a subsidiary of Technical Publishing Co. No part of this publication may be reproduced, stored in a retrieval system, or transmitted, in any form or by any means, electronic, mechanical, photocopying, recording, or otherwise, without the prior written permission of the publisher.

The views and opinions contained herein are those of the author and not necessarily those of the publisher.
ISBN 0-695-80677-7
Library of Congress Catalog #76-7181

CONTENTS

INTRODUCTION 4

chapter	page
1 CITIZENS BAND RADIO AND OTHER PERSONAL COMMUNICATIONS	7
2 A BRIEF HISTORY OF THE CITIZENS RADIO SERVICE	17
3 HOW TO BECOME A CBer	23
4 SELECTING A RIG	35
5 ANTENNAS FOR CITIZENS BAND	51
6 CB ACCESSORIES	69
7 INSTALLING YOUR NEW CB RIG	75
8 SECURITY	91
9 CB LINGO	103
10 SERVICE YOU CAN DO YOURSELF	117
11 CITIZENS BAND RADIO AND THE SPORTSMAN	125

chapter	page
12 THE FEDERAL COMMUNICATIONS COMMISSION AND YOU	133
13 CB CLUBS AND OTHER ORGANIZATIONS	143
14 AMATEUR RADIO	149
15 OTHER PERSONAL TWO-WAY RADIO MODES	161

appendixes	
A MANUFACTURER'S DIRECTORY	171
B FCC RULES AND REGULATIONS	175
FCC PERMANENT AND TEMPORARY LICENSE APPLICATIONS	187-190
C FCC MONITORING STATIONS	191
D FCC BUREAU OFFICES	192

INTRODUCTION

When we set out to write Citizens Band Radio Digest, we decided to produce a book that would be different from every other CB book on the marketplace. Our objective was two-fold: first, to provide the reader with a good basic grounding in CB technology and usage; and second, to introduce him to the other radio communications opportunities that are open to him in the United States.

Of course, we began our work with one big advantage over our predecessors—we'd seen what they'd done and hadn't done, and could pace our work accordingly. As a result, we think you'll find CITIZENS BAND RADIO DIGEST is more complete, more accurate, and more useful than any other CB book you've seen or are likely to see!

We make no claim that you'll find everything you'll ever want to know about CB or any other kind of radio in these pages. The complete book on any subject, much less one that has as many facets or is as volatile as the Citizens Radio Service, has never been written nor will it ever be. What you will find here is the information you'll need to get started in CB properly with respect to the FCC, your neighbors and your pocketbook.

CITIZENS BAND RADIO DIGEST isn't only a book for the beginner, however. Its pages contain a lot of information that the experienced CBer or even ham will find useful. For example, the discussions on specifications, installation and security contain some hints and kinks that are likely to be new to most two-way radio buffs, not only radio amateurs but professionals. And even if your interest in two-way radio is purely CB, don't pass up the chapters on the other phases of personal radio. They also have information and ideas you can apply to your CB activities, and who knows . . . you may even find they'll broaden your outlook and horizons as well!

To completely enjoy and make use of something, you need to know more about it than just the simple mechanics. That's why we've gone to a good deal of trouble to research the history of citizens radio and the FCC for you. Don't pass those chapters by—they, like the chapters on technical matters, can also give you insights that will contribute greatly to your CBing.

That, then, is pretty much what you'll find in the pages that follow. Enjoy them, profit by them, and then join us on the air. We'll be looking for you, Good Buddy!

chapter 1

CITIZENS BAND RADIO AND OTHER PERSONAL COMMUNICATIONS

you too can talk on the radio

Citizens Band radio was established by the Federal Communications Commission just after World War II to provide any citizen who wished it the personal use of two-way radio. Unlike business, marine or aircraft radio, each of which is dedicated to a specific purpose and can be used only for that specific purpose, citizens band radio is available to essentially any citizen who wants it with no questions asked as to why he wants it. There are legal limitations placed on the use of CB radio, to be sure, but they are relatively few and rather elemental. Amateur radio is like CB in that an amateur radio ("ham") operator has a great deal of freedom in who he can talk with and what he can talk about—far more than even the CBer. But before he can have that freedom, he has to pass a technical examination and demonstrate Morse code proficiency. We'll be talking more about the other "personal" radio services later in this book. For now, let's take a closer look at CB.

In terms of the number of licensees, the Citizens Radio Service is the largest radio service in the world. By the time the first buyer of this book reads these words, the Federal Communications Commission will have issued more than six million licenses for the CB Service, and with applications

(Facing page.) CB is useful in the outdoors as well as automobile and home. Here a hunter uses CB to alert a companion in another blind.

coming in at an average of about a half-million a month for the first half of 1976, the number of licensees could well double before these pages are a year old. Furthermore, few if any of those licenses are for a single transceiver. Most licensees end up with a base and at least one or two mobile units, though up to 15 transmitters may be licensed under a single call sign without special justification.

CB LICENSE APPLICATIONS
3½ year monthly growth

MONTH	1973	1974	1975	1976
JANUARY	23,000	32,000	73,000	513,000
FEBRUARY	20,000	35,000	62,000	472,000
MARCH	24,000	35,000	146,000	561,000
APRIL	24,000	35,000	173,000	488,000
MAY	25,000	43,000	162,000	449,000
JUNE	18,000	40,000	171,000	468,000
JULY	19,000	36,000	172,000	
AUGUST	19,000	36,000	193,000	
SEPTEMBER	13,000	38,000	216,000	
OCTOBER	18,000	42,000	286,000	
NOVEMBER	23,000	53,000	291,000	
DECEMBER	21,000	52,000	415,000	
TOTAL:	247,000	477,000	2,360,000	2,951,000 (half year total)

TV and popular music have both contributed to and profited from CB's astronomical growth. C.W. McCall's best selling record "Convoy" also landed him a promotional job with Midland, a leading CB and amateur radio equipment supplier.

In technical terms, almost all CB radio is operated on 23 channels (increasing to 40 channels January 1, 1977) located in what is known as the "11 meter band," a narrow portion of the radio frequency spectrum located roughly at the point separating the high frequency ("HF") from the very high frequency ("VHF") spectrums. These 23 channels range from 26.965 MHz (channel 1) to 27.255 MHz (channel 23), spaced 10 kHz apart, with the 17 new channels extending the top CB channel to 27.405 MHz. If some math sharpies noticed that there are more than 23 slots 10 kHz wide between 26.965 and 27.255 MHz, that is because there's another CB service—Class C—for radio controlled models sandwiched in between. We'll cover Class C as well as the original CB, Class A (in the UHF 460 MHz band) later on. Today's action is in Class D, and so are we—at least for the moment.

CB transmitters are limited by the rules to a maximum carrier power input of 5 watts and an output of 4 watts using conventional amplitude modulation (one of the more sophisticated single sideband [SSB] CB transmitters [discussed later in this book] can put out 12 watts *peak* power). This is not very much power compared to a small AM broadcast station which typically runs 1,000 watts *output* (the larger AM broadcast stations are licensed for 50,000 watts output) or the amateur service which is permitted 1,000 watts input, but under certain conditions even a 5-watt transmitter can reach half way around the world on 11 meters. Since the CB service is supposed to be for short range personal and business type communications, the FCC has placed a 150-mile limit on the range of permitted contacts, but when the skip* is in and signals from across the country (and beyond) are stronger than local stations a couple of miles away, the temptation to respond is more than a lot of people can withstand!

CB radio in a "social" sense has also become a way of life for many people. In many households the entire family is into CB, making friends with other families through radio contacts and attending CB club meetings and jamborees as a family group. Though an individual must be 18 or older to receive a CB license, any member of a licensee's immediate family can operate the "family" station—and usually does! Sociologists and psychologists have been doing a lot of studies of the CB "craze," and have developed theories that much of its popularity is the result of a current basic need to communicate with others without the dangers involved with face-to-face encounters. Perhaps so, but whatever the reason it has to be something more than Dad's desire to let his wife know he'll be a few minutes late for dinner that has gotten so many millions of Americans to put so much time and money into putting radio stations in their homes and cars!

"... short distance business or personal radio communications ..." is the purpose the FCC has assigned to the Citizens Radio Service. CB radio is used by a great many businesses for their operations—small taxi companies, garages and service stations, TV repair shops, delivery services, grocery stores and quick food shops all benefit from the advantages of inexpensive two-way radio communications. Many plants and factories also use CB radio—it does a very good job between a loading dock and a forklift truck in a storage yard, for example. For such

*Skip—Long-distance transmission of radio signals caused by radio waves bouncing off the ionosphere and reflecting back to earth.

> *"CB radio in a 'social' sense has also become a way of life for many people. In many households the entire family is into CB..."*

Though most CB communications involve automobiles and base stations, it's equally valuable for personal ship-to-shore or even private aircraft.

CB radio has actually been around for a long time, but it wasn't until the gas shortage and the resultant national 55mph limit that it became widely publicized. Now many travelers and most truckers —all but one in this lineup— are CB radio equipped.

users the CB radio has the advantages of a low cost and an easy-to-get license, against which must be weighed the problems of interference from the thousands of other users, both nearby and sometimes (when skip conditions are right) across the country.

By far the largest class of CB user is the private citizen, either for strictly personal reasons (car to home, boat to summer cottage) or as an indirect business tool (trucker or traveling salesman looking for advice on road conditions, for example). The use of CB by travelers has received the most publicity, starting with the truckers' strike and gasoline shortage a couple of years ago which greatly accelerated public awareness and interest in CB radio. Now, of course, CB has become the subject of a number of popular songs and sees frequent use in TV dramas as well as TV and radio commercials.

CB is truly a valuable traveling companion, for both the professional cross-country trucker and for the family off for a weekend in the country. Away from the larger cities channel 19 (the informally agreed on "highway channel") is usually continually busy with road-related chatter. "Smokey reports" are always prominent, of course, but for those who don't feel the need to exceed the "double nickel" (55 mph speed limit), the exchange of information on lane closings, short cuts, best gas prices or even eating places makes a CB set a good investment for a cross-country trip. When trouble strikes, a CB unit can be priceless. With so many CB-equipped cars and trucks on

CBers and hams alike provide emergency and public service communications for their communities.

the road today a call for assistance after a breakdown or accident will not only bring an immediate response but—relayed from traveler to traveler if necessary—a specific promise of aid from the nearest service area.

For truly emergency situations channel 9 is the place to go. It has been designated by the FCC for only "Emergency communications involving the immediate safety of life of individuals or the immediate protection of property or... Communications necessary to render assistance to a motorist." Channel 9 is monitored not only by many highway patrol and police departments but also by several volunteer groups such as REACT (Radio Emergency Associated Citizens Teams) and ALERT (Affiliated League of Emergency Radio Teams) on a countrywide 24-hour-a-day basis. Well-trained REACT or ALERT volunteers respond to a motorist's call for help, call whatever agency can best offer him direct assistance, and then stand by until the stranded traveler advises that help has actually arrived. The volunteer monitors also do a very good job of keeping channel 9 clear of careless channel switch flippers so the channel can be available at all times for its designated purpose.

CB radio sees a lot of point-to-point (base station to base station) use, too. In city areas this takes the form of casual ham radio type "rag chewing," an activity the FCC frowns on. In areas away from civilization CB radios can provide badly needed communications which otherwise wouldn't exist. Isolated farm buildings are often linked

> *"When trouble strikes, a CB unit can be priceless. With so many CB-equipped cars and trucks on the road today a call for assistance will bring an immediate response..."*

Marine radio is an important personal communications mode and a must for most boaters.

together by CB radio as are summer cottages, public and private recreation facilities, construction sites—and of course lonely housewives and teenagers with homework problems, just like in the days of the old fashioned telephone "party lines."

Marine CB is popular in many boating areas, though marine VHF/FM (which we'll discuss in detail later on) is usually a far better choice for the boater. The small boat enthusiast will find CB very good for communicating with other boaters interested in fishing or with his family back on shore at the cottage or automobile. However, CB is not in the same league with VHF/FM when it comes to safety and emergency situations. Channel crowding and improper operation have so limited CB in the eyes of the Coast Guard that they have flatly refused to have anything to do with CB! However, the Coast Guard does recognize the fact that boats equipped only with CB can get into trouble just like any others and will respond to emergency calls relayed to them by CB shore stations. For the boater who spends all or most of his time on smaller bodies of water or in other areas well removed from Coast Guard facilities, CB is usually the only means of ship-to-shore communications open to him.

Citizens Band provides what is supposed to be but not always is a short range means of personal communications. Under *ideal* conditions a pair of well equipped, legal (4 watts output) CB base stations should be able to communicate well at a distance of 25 miles or so. Under the same conditions base-to-mobile range will be on the order of 10 miles, while satisfactory mobile-to-mobile range drops down to 5 miles or so. Unfortunately, few CB contacts take place under ideal conditions since CB's popularity and resultant channel crowding practically insures that city-bound CBers will find someone talking on any given channel at almost any hour of the night or day. In addition, automotive ignition and other electrical noise sources can and frequently do drown out an otherwise readable signal, particularly in mobile operation.

Interference from distant (skip) stations is another problem that occurs on the 11 meter band. Skip conditions occur most frequently on 27 MHz at times of high sunspot activity, and sunspot activity varies on an 11-year cycle.

CB's phenomenal popularity has resulted in tremendous growth for the industry. This JAL 747 jumbo cargo jet brought nearly 100 tons of electronic components from Japan to Hy-Gain to be used in the manufacture of CB radios.

The tremendous spurt in CB growth during 1975 and 1976 came at the very low point of the present sunspot cycle, so most current CBers know skip more as a novelty than as a plague. But FCC Chief Engineer, Ray Spence, predicted in a recent speech that the next period of high sunspot count—due in a few years—would cause such severe skip interference to 11 meter CB that mobile-to-mobile range on city streets would be effectively reduced to one mile during most daylight hours! However, even if the sunspot problem does become that severe, CB isn't going to disappear—or even diminish. The benefits—and fun—of CB radio are too great to be turned off by increasing interference!

Though flight instruments make up the heart of this Cessna Cardinal's instrument panel, more than half (including the entire right-hand side) is devoted to radio or radio-navigation.

CHANNEL NUMBERING SYSTEM
27 MHz CITIZENS RADIO SERVICE

Frequency	Class D (Old)	Class D (New)	Class C*	Frequency	Class D (Old)	Class D (New)	Class C*
26.965 MHz	1	1		27.195 MHz			RC*
26.975 MHz	2	2		27.205 MHz	20	20	
26.985 MHz	3	3		27.215 MHz	21	21	
26.995 MHz			RC*	27.225 MHz	22	22	
27.005 MHz	4	4		27.235 MHz		24**	
27.015 MHz	5	5		27.245 MHz		25**	
27.025 MHz	6	6		27.255 MHz	23	23**	
27.035 MHz	7	7		27.265 MHz		26	
27.045 MHz			RC*	27.275 MHz		27	
27.055 MHz	8	8		27.285 MHz		28	
27.065 MHz	9	9		27.295 MHz		29	
27.075 MHz	10	10		27.305 MHz		30	
27.085 MHz	11	11		27.315 MHz		31	
27.095 MHz			RC*	27.325 MHz		32	
27.105 MHz	12	12		27.335 MHz		33	
27.115 MHz	13	13		27.345 MHz		34	
27.125 MHz	14	14		27.355 MHz		35	
27.135 MHz	15	15		27.365 MHz		36	
27.145 MHz			RC*	27.375 MHz		37	
27.155 MHz	16	16		27.385 MHz		38	
27.165 MHz	17	17		27.395 MHz		39	
27.175 MHz	18	18		27.405 MHz		40	
27.185 MHz	19	19					

*RC Class C Channels for Radio Control.
**Note Change in Sequence.

chapter 2

A BRIEF HISTORY OF THE CITIZENS RADIO SERVICE

how we got here from nowhere and where we may be tomorrow

CB radio is something of a war baby—World War II, that is. Two-way radio technology made tremendous strides during the war years, of course, and one of the wartime FCC commissioners, E. K. "Jack" Jett, was intrigued by the idea of applying the new technology to use by the everyday citizen. Commissioner Jett's idea became a reality in 1947 when the FCC assigned the 460-470 MHz band to the new Citizens Radio Service. Two classes of licenses were established for the new service: Class A, which used relatively sophisticated and expensive equipment and was appropriate for the professional (businessman) user; and Class B, which used rather crude (very crude, by today's standards) and inexpensive equipment and was for the general public. Unfortunately the technology needed to make good performing radios at reasonable cost for UHF frequencies didn't exist at that time. That, combined with propagation characteristics that still restrict transmission range at those frequencies, left the relatively few users dissatisfied and resulted in little interest in citizens radio for the first 10 years or so of its existence.

However, the FCC was still interested in making personal two-way radio more attractive to more people, so in

(Facing page.) CB really came into its own when the 27 MHz band was opened in 1958 offering greatly improved performance at reasonable cost. Tube-equipped rigs prevailed for more than a decade, and though they are bulky and comparatively power hungry, many better quality older rigs like this mid '60s Cobra are still providing good performance in CB base stations.

The first citizens band transceiver to be type accepted by the FCC for use in the just established 465 MHz band. It was made by Citizens Radio Corporation, Cleveland, Ohio, and announced to the public on December 14, 1948.

ORIGINAL CB FREQUENCIES (1947)	
Class A	**Class B**
460-462 MHz	462-468 MHz
468-470 MHz	

CB FREQUENCIES TODAY (January 1, 1977)		
Class A	**Class C**	**Class D**
462.550-462.725 MHz	26.995, 27.045, 27.095	26.965-27.405 MHz
467.550-467.725 MHz	27.145, 27.195 and 27.255 MHz	(40 channels)
	72.16, 72.32, 72.96; 72.08, 72.24	
	72.40 and 75.64 MHz*	

*72.08, 72.24, 72.40 and 75.64 may be used only for controlling model aircraft

Current mobile rigs are more compact and designed to fit in with a modern auto's decor. This Motorola Mocat is designed for under-dash mounting.

1958 they reallocated part of the 11 meter band (27 MHz) to the Citizens Radio Service. This band of frequencies, which had been shared by amateur radio and a variety of special industrial users, was divided into 29 channels spaced 10 kHz apart for use by the Citizens Radio Service. Six of the new channels were for use in radio control applications such as garage door openers and radio controlled models (Class C Citizens Band) while the remaining 23 channels were to be exclusively for voice communications. At the same time the original UHF/CB frequencies were drastically reduced, first to 49 channels and later to the present 16 (which also ended the issuing of Class B licenses so that that class went out of existence when the last Class B license expired).

The move to 27 MHz made all the difference in the world to CB communications. Not only did low cost equipment with excellent performance become possible, but signal propagation at the much lower frequency was such that reliable communications over distances of 20 miles or more was readily achieved. A number of manufacturers began building CB equipment, and the service started its first real growth.

The new band was not without its problems, however. Along with the increase in businesses and individuals using CB radio for appropriate purposes, the convenience of mobile-to-mobile and mobile-to-base communications attracted a new group of CBers, those who wanted to communicate for communication's sake much in the fashion of ham radio operators. Coupled to this was the fact that 27

Today's base station is compact yet offers features and performance unheard of a decade ago. This Pearce-Simpson Bengal *can be used on both AM and single sideband.*

MHz is the top of the shortwave spectrum so, under certain fairly common conditions, it can propagate signals over distances of a thousand miles or more. Since hobby type and long distance ("DX") communicating was not what the FCC wanted or expected of CB radio, rules were put into effect defining the types of communications that were permitted and limiting the distance over which CBers could legally communicate to 150 miles.

Human nature being what it is, and the number of CB operators vastly exceeding the number of FCC field engineers available to enforce the rules, the CB "hobby" grew far more rapidly than the CB *service*, and the foundations for the present channel "crunch" were laid. Since use of FCC call letters by a CB hobbyist was likely to lead to a "pink ticket" (FCC citation), the hobbyists began using "handles" to conceal their identities. And with call letters going unused, a license wasn't even necessary! The next step was obvious. As hobby CBing became dominant, the competition pushed hobbyists into adding illegal power amplifiers to their transceivers or using modified and much more powerful amateur transmitters—all into illegally high antennas. Citizens Band radio was out of control!

In 1975 the FCC came to grips with the problem it had created by putting into effect several rules changes and proposing some others. Restrictions on permitted communications and use of call signs were relaxed in such a way as to encourage proper operation, antenna height restrictions were liberalized, and the manufacture and sale of 11 meter power amplifiers was strictly prohibited. Ex-

Vocaline JRC-400, probably the best known of the early (mid 1950s) commercial equipment for UHF/CB. Designed for mobile or fixed Class B use, the JRC-400 sold for under $140.00 a pair while Class A operation required commercial equipment selling for many times that.

Stewart Warner Portafone was an early hand-held CB. The Portafone was also a Class B unit and sold for about $200 a pair. Stewart Warner also made 6 and 12 VDC and 115 VAC power packs to operate the Portafone as a mobile or base station.

pansion of the number of 27 MHz CB channels was also proposed in 1975 and—as a result of rule making by the FCC in July, 1976—the original 23 Class D channels were increased to 40 channels effective January 1, 1977. In addition, a new Class E citizens band to be located in the present amateur radio allocation at 220 MHz has been pending for several years. This does not appear likely to happen at the frequency proposed because of potential television interference and amateur radio opposition. It does look very likely that sometime in the next year or so a new UHF or VHF/CB band with many more channels will be allocated to the Citizens Radio Service, the largest and most rapidly growing in the world. In early 1976 a temporary license system was established to permit new applicants to get on the air legally immediately after applying for a permanent license, removing one of the most powerful incentives for newcomers to start out operating outside of the rules.

That the Citizens Radio Service will continue to grow is a certainty, and with 6 million or 12 million or even more licensees there is no question that the present 11 meter band—even with the expansion to 40 channels—could not properly serve them. New CB channels in the VHF, UHF or even microwave spectrums will be forthcoming if not this year within the next year or so. U.S. CB user groups and manufacturers have presented a request to the FCC that it propose worldwide recognition for the Citizens Radio Service at the forthcoming World Administrative Radio Conference (WARC). This conference, scheduled for 1979, is a meeting of representatives of all the world's nations at which their radio needs will be examined and frequencies for those needs allocated on a worldwide basis. A few other nations such as Canada already have an official CB service comparable to that of the U.S. operating on the 27 MHz band, while unofficial and/or illegal CB operation exists in many other parts of the globe. The U.S. WARC CB proposal would not only establish the 27 MHz band on an international basis, but would ask for an additional frequency assignment for the service in the same portion of the VHF spectrum specified in the request for the new Class E CB band that is still pending before the FCC.

Whenever and wherever citizens radio expands, it will not drastically change things for present day CB users. No one really knows how many Class D (27 MHz) CB radios are presently in use in the United States, but estimates of 15 to 20 million sets seem as valid as any. That many radios on only 23 or even 40 channels is a big part of the

CB problem, but it also contributes a large part of CB's value when trouble strikes. Someone nearby is listening and *is* going to hear your call for help! Though changes in CB will be taking place—next year or in 1979—now is still the time to get into CB radio. Despite its problems CB radio is plenty of fun and a valuable tool—a tool that could even save your life!

What the next generation of CB equipment will look like is anyone's guess, with SBE proposing keyboard entry channel selection for both base and mobile as in these designer's concepts.

chapter 3

HOW TO BECOME A CBer

come on "good buddy," the copy's fine!

Before getting into the "how" of your becoming a CBer, perhaps a short review of the "why" you should become part of the CB boom is in order. The "why" is actually spelled out in the first paragraph of the Federal Communications Commission's Rules governing the Citizen's Radio Service, paragraph 95.1 (Basis and Purpose). It says that CB radio is "to provide for private short-distance radio communications service for the business or personal activities of licensees . . ." and that pretty well sums it up. If you've ever wished you could talk from car to house or office when you're hung up in traffic, ask a motorist coming toward you on the highway what road conditions ahead of you are like, or check with a campground for a vacancy while still on the interstate, CB radio's for you!

    The very first thing you should do is learn something about CB. You've already made a good start—you're reading this book! However, no one ever learned to ride a bicycle by reading a book, and none of the dozens of CB books on the market can tell you everything you'll want to know about where you fit into personal two-way radio communications. For that you'll want to do some face-to-face two-way communicating with friends who are already

(Facing page.) If you've ever wished you could talk from car to house or office or ask a motorist coming toward you what road conditions ahead of you are like, CB radio's for you!

CB radio gives the user a "handle on the world"—a means of communication that would otherwise be difficult or impossible.

CB is a tool for many users. This farmer can call in for supplies or assistance without having to make a trip in.

Truckers make good use of their CBs, not only for road, fuel and food tips but also to keep each other awake on late night hauls.

into CB, hams (if you know any) and the (hopefully) friendly people at your nearby CB store.

All three sources have limitations, however. Your CBing friends are, if they're active and really into CB, the ones who will know what's going on in your area, both on and off the air. They can take you to club meetings and jamborees, tell you which channels are best locally for what type of activities, and steer you to stores where you'll get genuine help in getting started in CB instead of being pressured into buying a high-priced rig with features you'll never use. However, most CBers are people like you with little or no technical background so—with a few rare exceptions—they usually aren't well qualified to offer sound technical advice. Unfortunately many don't know this. Any evening's listening to the CB chatter in any large city can bring you the most amazing rewrites of the laws of physics, electricity and even common sense! Don't flatly reject your CBing friends' technical advice—it may be valid and valuable—but do be skeptical of it.

Hams are a much better source of technical assistance. Though some hams have become "appliance operators," they all still had to pass a fairly stiff technical exam to get started. Most active hams can be pretty helpful when it comes to such things as antenna installation, power wiring, electrical noise problems and the like. Don't expect a ham friend to recommend a CB radio or antenna to you, though. Unless he is into CB himself, he'll probably know as little about what's on the current CB marketplace as the average housewife does about power lawn mowers. That isn't to say a ham can't be helpful in comparing specifications or evaluating how a rig performs or an antenna is built. Just don't expect him to know the good and bad points of the dozens of CB radios or accessories on the current marketplace.

One problem you may encounter when asking a ham to help you with CB is that some hams deeply resent CB. Before Class D CB came along, the 27 MHz CB band was a ham band, and though it was not heavily used, the hams that did operate there liked it very much. Despite the loss of 11 meters by Amateur Radio to CB having occurred almost 20 years ago, it still hasn't been forgotten. In addition the hobbyist CBer also irritates those hams who feel that if it wasn't for CB, the CB hobbyist would have mustered a little more ambition, learned Morse code and theory, and become a ham. Don't be surprised if a ham friend who hears you are thinking CB starts working on you to become a ham—and don't reject his efforts, either!

You can get a lot of help getting into CB from the people at your local CB equipment outlet. CB specialty shops and amateur radio stores that also handle CB gear are your best bet—their people are almost invariably technically competent, and their philosophy is that a satisfied customer is a repeat customer. The electronic chain stores can be very good or terrible, depending on how lucky they've been finding a competent CBer for a counterman. The same thing is true of the big retail outlets and discount houses, so you'll have to get friends' advice as to where to go to get the best shake in your area. Wherever you go for help and advice once you've decided that you do want to join us on 27 MHz, you'll need a license. Let's talk about that.

GETTING YOUR LICENSE

There's no problem involved in getting your CB license, and it doesn't cost much, either. All you need to do is to fill out the application (we've included one, FCC Form 505, in the back of this book so you won't even have to move out of your chair to get one) and mail it with your check or money order for $4 to the FCC's license processing facility in Gettysburg, Pennsylvania. (Your personal check is the best, by the way. If there is any foul up in processing your license, the cancelled check is proof you paid for it!). Until early 1976 that was it—then you waited . . . and waited . . . and waited until your license finally arrived. The problem was that Gettysburg had become literally buried under the tidal wave of CB license applications that started building up a couple of years ago. The result—unhappy applicants were waiting 4 or 5 months for their licenses.

> *"There's no problem involved in getting your CB license, and it doesn't cost much, either."*

For many the answer was to choose a nice "grabby" handle and get on the air anyway—patience lasted only so long and besides, they had plenty of unlicensed company. The FCC wasn't ignoring the problem, however. Not only did they do some fantastic things in speeding up processing, but they also introduced a new "instant" temporary license that permitted license applicants to get on the air immediately, legally! Naturally, we've put one of those—FCC Form 555-B—in the back of this book for you, too.

FCC's new "instant" license is strictly temporary, however. It expires 60 days from the day you fill it out. That's no problem now for most people since the staff at the Gettysburg processing facility now has CB licenses back to applicants' mailboxes a month or so after their applications go in the mail. Note we said *most* people, because that's where the rub lies—make a mistake on your Form 505 or forget your check, and your temporary license will expire for sure before your permanent ticket gets issued. Now the FCC isn't likely to take anyone to court for using a temporary call sign beyond the prescribed 60-day period, but it's still a lot easier for everyone if your application is done right the first time and you don't get snarled in any red tape. Just for the record, let's go through the application and temporary license step by step.

FILLING OUT FORM 505

Take a look at the sample form we've made out as an example. First note the nice, clear, *readable* printing on our example. If the FCC people can't read your application, they can't process it. And if they read it incorrectly, it can be just as bad. The FCC gets hundreds of licenses back from the post office each week as undeliverable, primarily a result of misread applications. If possible, use a typewriter. Whatever you do, make it readable!

Item 1—your name—seems pretty obvious, but just be sure you do it (a) first name, (b) middle initial, (c) last name—not last name first.

Item 2 causes a lot of rejects—it's supposed to be your birthday, but thousands of would-be CBers have filled in today's date. Don't be one of them!

Item 3 is for business applicants. If you're applying for a license for a business, put the business name here. If you're not, skip it.

Items 4A, 4B, 4C and 4D are for the mailing address of the person or organization applying for the license.

Items 5A, 5B and 5C are needed only if item 4 is a P.O. Box or RFD number. If they are, the FCC wants to know where the principal station of the licensee will actually be located.

Item 6 is to indicate what you are—an individual, a corporation, etc. Check the correct box.

Item 7 asks whether you're a new applicant, an applicant renewing an existing license or if you wish to increase the number of transmitters you operate. If you're reading this, the chances are you're new. Check the appropriate box.

Item 8 lets you choose between CB Class C (non-voice-radio control for models) and Class D (voice communications—what you want). Put your "X" in the box beside Class D.

> "The FCC... did some fantastic things in speeding up processing, (and) introduced a new "instant" temporary license that permits license applicants to get on the air immediately, legally!"

FCC FORM 505

August 1975

INSTRUCTIONS

A. Print clearly in capital letters or use a typewriter. Put one letter or number per box. Skip a box where a space would normally appear.
B. Enclose appropriate fee with application. Make check or money order payable to Federal Communications Commission. DO NOT SEND CASH. No fee is required of governmental entities. For additional fee details see FCC Form 76-K, or Subpart G of Part 1 of the FCC Rules and Regulations, or you may call any FCC Field Office.
C. Mail application to Federal Communications Commission, Gettysburg, Pa. 17326

United States of America
Federal Communications Commission

Form Approved
GAO No. B-180227(R01 02)

APPLICATION FOR CLASS C OR D STATION LICENSE IN THE CITIZENS RADIO SERVICE

NOTICE TO INDIVIDUALS REQUIRED BY PRIVACY ACT OF 1974

Sections 301, 303 and 308 of the Communications Act of 1934 and any amendments thereto (licensing powers) authorize the FCC to request the information on this application. The purpose of the information is to determine your eligibility for a license. The information will be used by FCC staff to evaluate the application, to determine station location, to provide information for enforcement and rulemaking proceedings and to maintain a current inventory of licensees. No license can be granted unless all information requested is provided.

1. Complete ONLY if license is for an Individual or Individual Doing Business AS

FIRST NAME: JOHN
INIT: L
LAST NAME: SAMPLE

2. DATE OF BIRTH
MONTH: 5 DAY: 12 YEAR: 48

3. Complete ONLY if license is for a business, an organization, or Individual Doing Business AS
NAME OF BUSINESS OR ORGANIZATION:

4. Mailing Address
4A. NUMBER AND STREET: 420 MAINE
4B. CITY: ANYTOWN
4C. STATE: IL
4D. ZIP CODE: 66666

NOTE: Do not operate until you have your own license. Use of any call sign not your own is prohibited

5. If you gave a P.O. Box No., RFD No., or General Delivery in Item 4A, you must also answer items 5A, 5B, and 5C.
5A. NUMBER AND STREET WHERE YOU OR YOUR PRINCIPLE STATION CAN BE FOUND (If your location can not be described by number and street, give other description, such as, on RT. 2, 3 mi., north of York.)
5B. CITY:
5C. STATE:

6. Type of Applicant (Check Only One Box)
☒ Individual
☐ Business Partnership
☐ Sole Proprietor or Individual/Doing Business As
☐ Association
☐ Governmental Entity
☐ Corporation
☐ Other (Specify)

7. This application is for
☒ New License
☐ Renewal
☐ Increase in Number of Transmitters

IMPORTANT Give Official FCC Call Sign

8. This application is for (Check Only One Box)
☐ Class C Station License (NON-VOICE—REMOTE CONTROL OF MODELS)
☒ Class D Station License (VOICE)

9. Indicate number of transmitters applicant will operate during the five year license period (Check Only One Box)
☒ 1 to 5 ☐ 6 to 15 ☐ 16 or more (Specify No. and attach statement justifying need.)

10. Certification   I certify that:
- The applicant is not a foreign government or a representative thereof.
- The applicant has or has ordered a current copy of Part 95 of the Commission's rules governing the Citizens Radio Service. See reverse side for ordering information.
- The applicant will operate his transmitter in full compliance with the applicable law and current rules of the FCC and that his station will not be used for any purpose contrary to Federal, State, or local law or with greater power than authorized.
- The applicant waives any claim against the regulatory power of the United States relative to the use of a particular frequency or the use of the medium of transmission of radio waves because of any such previous use, whether licensed or unlicensed.

WILLFUL FALSE STATEMENTS MADE ON THIS FORM OR ATTACHMENTS ARE PUNISHABLE BY FINE AND IMPRISONMENT. U.S. CODE, TITLE 18, SECTION 1001.

11. *John L. Sample*
Signature of: Individual applicant, partner, or authorized person on behalf of a governmental entity, or an officer of a corporation or association

12. Date 10/11/76

Item 9 is the place where you specify the maximum number of transmitters you intend to operate under your license. Most CBers, even those with two cars and a boat, are unlikely to need more than five transmitters. Unless you really think you'll want more, check the 1 to 5 box.

Item 10 is your certification that you are and will be a law abiding CB operator as attested by your signature in item 11.

Items 11 and 12 are your signature and the date of application. Of all the errors that can be made on Form 505, failure to sign and/or date the application is the most common. Don't overlook these.

Now that you've finished Form 505 you're still not done. The first item of business is a properly addressed envelope. Note that CB applications go to the FCC at Gettysburg, Pennsylvania 17326—not 17325, the zip code the rest of the FCC's mail goes to. That's because they have received so much more CB mail than anything else that the post office gave them a separate zip code to help keep it sorted out. Next remember the fee—$4 check or money order, please. Don't forget to sign the check, and write it on a solvent bank account. It's surprising how many of each month's half million or so CB applications are paid for by bum checks, a practice Uncle Charlie gets very unhappy with. Double check.

- Form 505, completely and properly filled out, signed and checked.
- Check or money order made out to Federal Communications Commission.
- Envelope addressed to Federal Communications Commission, Gettysburg, Pennsylvania 17326.
- First class postage.

Put them all together, drop the result in the outgoing mail, and you've completed phase one of your licensing program.

YOUR INSTANT CB LICENSE

Once you've completed and mailed your Form 505 application for a CB license, you're ready to fill out your Temporary Permit and get on the air—legally. The temporary permit—FCC Form 555-B—is for your own use, but in the off chance that you might run afoul of a FCC inspector or some other government official, you'd better fill it out before you go on the air.

Form 555-B is simplicity itself. Simply read the instructions Part (1), check all the boxes, put your name, address

CITIZENS BAND RADIO DIGEST 29

FCC FORM 555–B
April 1976

United States of America
Federal Communications Commission

Temporary Permit
Class D Citizens Radio Station

1 Instructions

- Use this form only if you want a temporary permit while your regular application, FCC Form 505, is being processed by the FCC.
- Do not use this form if you already have a Class D license.
- Do not use this form when renewing your Class D license.

2 Certification
Read, Fill In Blanks, and Sign

I Hereby Certify:

☒ I am at least 18 years of age.
☒ I am not a representative of a foreign government.
☒ I have applied for a Class D Citizens Radio Station License by mailing a completed Form 505 and $4.00 filing fee to the Federal Communications Commission, Gettysburg, PA. 17326.
☒ I have not been denied a license or had my license revoked by the FCC.
☒ I am not the subject of any other legal action concerning the operation of a radio station.

Name: JOHN L. SAMPLE
Signature: John L. Sample
Address: 420 MAINE Anytown, IL
Date Form 505 mailed to FCC: 10/11/76

If you cannot certify to the above, you are not eligible for a temporary permit.
Willful false statements void this permit and are punishable by fine and/or imprisonment.

3 Temporary Call Sign

- Complete the blocks as indicated.
Use this temporary call sign until given a call sign by the Federal Communications Commission.

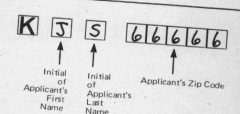

K J S 6 6 6 6 6

Initial of Applicant's First Name | Initial of Applicant's Last Name | Applicant's Zip Code

4 Limitations

Your authority under this permit is subject to all applicable laws, treaties and regulations and is subject to the right of use or control by the Government of the United States.

This permit is valid for 60 days from the date the Form 505 is mailed to the FCC.

You must have a temporary permit or a license from the FCC to operate your Citizens Band radio transmitter.

Do Not Mail this form, it is your Temporary Permit.

See the reverse side of this form for a summary of operating instructions.

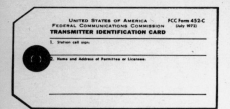

There's only one more thing you have to do before you can get on the air legally—identify your transmitter. FCC Rules require that every transmitter have an identification notice attached to it that shows its call letters and the name and address of the licensee. They even supply a tag—FCC Form 452-C—for the purpose, but you can do the job just fine with a piece of masking tape.

A well equipped full service store offering CB equipment is a priceless aid to the beginning CBer. Extensive inventory plus knowledgeable, helpful countermen are signs that you've come to the right place.

and date you mailed Form 505 on the proper lines and sign Part(2) on the signature line. For Part(3), which will give you your temporary call sign, put your first and last initials in the right squares, add your ZIP code where it belongs, and—Congratulations!—you're now a licensed CB operator, complete with call letters!

BUYING YOUR FIRST RIG

What you're going to want in a CB rig is pretty well determined by what you want to do with it. Of course, the differences between an AC line-powered base station and a compact mobile radio designed for 12 VDC operation are obvious, but most base rigs have provisions to be operated from a car battery for mobile (or emergency) use, and a mobile radio makes a very nice compact base station when operated from an inexpensive regulated 12 VDC power supply.

For starting out you may well want to consider buying a less expensive new set without a lot of fancy features, or even a second-hand radio. One thing is for sure—you will want and need advice you can trust. A CB dealer who is respected by CBers in your area is one of the very best places to go for assistance in choosing your first rig. He has a stake in helping you to make a good first choice since you're a potential repeat customer. If you like what he does for you with your first purchase, you'll be coming back to him when you're ready to upgrade.

Used gear can be tricky, and unless you really know what you're doing or have *competent* expert advice, you're much better off buying a used radio from a reputable dealer. Radios bought at garage sales or flea markets are often bad news, either "sick" or hot—the bigger the bargain the more doubtful its parentage! The advantage of the dealer-bought used radio is that it's probably a trade-in from an upgrading customer and the dealer will stand behind it—at least to some degree—if you have problems.

Don't buy a used antenna. Weather gets into hardware and cables and causes no end of problems. New antennas aren't all that expensive, so ask your dealer (or that reliable friend) to recommend an inexpensive basic antenna to meet your needs.

We'll be covering rigs and antennas in much greater detail in the succeeding chapters. As a beginner in CB, solicit advice wherever you can—but maintain a critical ear so you can learn to know the difference between good advice and bad advice!

CB jamborees are both social occasions and flea markets. Good deals on equipment and accessories are generally available, but be cautious about "hot" or non-working gear.

OPERATING PRACTICES

There's no substitute for experience when actually getting on the air—and there's no way to get that experience other than getting on the air. At first what you hear will be confusing at best, but try to remember the people you hear chattering away so glibly were once just as green as you. Listen in to what's being said and note what channel it's being said on. Though channel 9 is the universal emergency channel (stay off 9 unless you're having an emergency or can help someone else with one) and channel 19 is invariably truckers and motorists talking road information, you'll find most of the other channels used by different specific groups in different parts of the country.

After you've done some listening, you'll probably find you're understanding more of what you're hearing. Study our chapter on CB lingo—it'll clue you in on many of those weird CB terms plus 10 codes and Q signals. Now—hopefully—you may have even found some people talking whom you think you'd feel comfortable talking with. You've got the rig and the license . . . talk to them! Best bet is to select a good strong signal for your first contact since he's most likely to hear you and you'll be less prone to getting "stepped on" (wiped out by interference).

"Breaker channel 6!" is the way to announce yourself, and "Go ahead breaker!" is the response you'll get. Come back, tell your new friend where you are, your "handle" (if you've selected one) and that you're brand new on CB. He'll undoubtedly welcome you, give you a signal report and introduce you to any friends who may be standing by on the channel. Congratulations! You've actually made a two-way radio contact from your own station!

" 'Breaker channel 6!' is the way to announce yourself, and 'Go ahead breaker!' is the response you'll get."

> **MAKING A CONTACT**
>
> It's all well and good to discuss the philosophy and psychology of CB, the procedure to follow to get a license and the mechanics of choosing and installing a radio, but when the moment of truth arrives and you're sitting with a microphone clenched in your fist what do you do next? Why, push that mike button and say:
>
> *You:* "Breaker channel 12!"
>
> *Him:* "Go ahead, breaker."
>
> *You:* "Thanks. This is 'Loan Shark,' KAB - 1234, looking for a signal report."
>
> *Him:* "10-4,* Loan Shark, you're 10-2* copy at Main and Third. What's your '20'*?"
>
> *You:* "Just south of city hall on Central. What's your handle?"
>
> *Him:* " 'Road Runner' here, Loan Shark. Just got to the home '20 so thanks for jawing with us. Road Runner, KXY - 6789 down.*"
>
> *You:* "Thanks, Road Runner. See you on the flip-flop*. Loan Shark, KAB - 1234 on the side.*"
>
> * 10-4: OK or message received
> * 10-2: Loud and clear or solid copy
> * '20: Location or 10-20
> * down: off, clear and shutting down; leaving the air
> * flip-flop: return trip
> * on the side: standing by

"It has become an accepted practice in many areas to simply break into an existing conversation and ask if you can make a call. Most operators co-operate courteously with such a request..."

Don't forget, however . . . you've got a couple of legal obligations to take care of. The most important of these is identification. When you sign clear of another station, the rules require that you must sign your call sign. Most operators combine their handles and call signs in their signoff as: "This is 'Carrot-Top,' KAB-01234, on the side." That's all it takes and you're legal (Note, by the way, Carrot-Top's call sign—its five digit number indicates he's still operating with his FCC Form 555-B Temporary Permit while waiting arrival of his permanent license).

The other operating practice you're supposed to observe is the "5 minutes on, 1 minute off" rule. What this rule means is that after 5 minutes of continued two-way conversation you (and the station or stations you're communicating with) are supposed to go off the air for at least one full minute. The purpose of the rule is to give other operators a chance to use the channel for their communications. It's a good rule since it gives an operator who needs the channel but is reluctant to break in on someone else's conversation a much needed opportunity. However, it has become an accepted practice in many areas to simply break into an existing conversation and ask if you can make a call. Most operators co-operate courteously with such a request, providing the breaking station moves

GOOD CB PRACTICES
(adopted from a recent FCC Public Notice)

1. *Channel Selection*—In selecting a channel for your station, it is very important that the following factors be considered:

    a.) There are a total of 40 channels available to all CB stations on a *shared basis*;

    b.) Channel 9 may be used *only* for communications involving immediate or potential emergency situations and/or assistance to motorists;

    c.) You may select any of the remaining 39 channels to conduct your normal personal or business radio communications;

    d.) To prevent unintentional "bleed over" interference to channel 9, the FCC recommends that all transmissions involving highway travelers be conducted on channels several channels removed from channel 9.

2. *Channel Useage*—Cooperate with all CBers to the fullest extent possible in sharing the CB channels—always try to be courteous and considerate when using a channel. In order to assure that all CB operators will have an equal opportunity to use the frequencies, radio communications between CB stations (interstation) must be limited to no longer than 5 continuous minutes to be followed by a silent period of at least 1 minute. Every CBer should restrict their time on the air to a practical minimum. We cannot stress enough the importance for all CBers to discipline themselves from needlessly transmitting for long periods of time.

3. *Identification*—Identify your radio transmissions with your own FCC issued call sign before and after each transmission. This call sign is unique in that it is unlike any other CB radio station call sign. Be proud to identify your radio transmissions with it. "Nicknames" or "handles" may also be used to identify your radio transmissions *provided they are accompanied by the FCC assigned call sign*. It is not necessary to transmit the call sign of the station with whom you are talking.

4. *Equipment*—Have frequency, power and modulation measurements made at regular intervals. Do not tamper with the equipment. A licensed commercial operator is required for any adjustments that might affect the proper operation of the station.

5. *Keep Informed*—You may obtain the FCC Rules and Regulations (Volume VI, Part 95) governing the Citizens Radio Service from the Superintendent of Documents, Government Printing Office, Washington, D.C. 20402. Copies of these Rules are also available at many radio equipment stores at a nominal cost, and after January 1, 1977 will be supplied by manufacturers with all new radios.

6. *Promote "Good CB Practices"*—Encourage other CB users to follow the above suggested practices.

If all CB users make a serious attempt to understand and follow the above recommended practices, we believe efficient utilization of the 40 shared CB channels will be maximized.

---

his traffic* to another channel if he raises the station he's looking for and if the traffic is going to be more than a short, "I'll be home in half an hour."

That pretty well covers the how and why and when of getting licensed and on the air. Now we're going to take a closer look at the hardware you're going to be using.

A brief summary of good CB practices. Note: A complete copy of the 1977 FCC Rules and Regulations are included in the back of this book.

*Traffic—Conversation or messages

chapter 4

SELECTING A RIG

it's not only what you say but what you say it on!

CB radios can be classified in three broad categories: base stations, mobiles and hand-held. Within each classification you'll find a wide variety—plain ones, fancy ones, big ones, little ones, good ones and not so good ones. That very variety is going to make it pretty tough when you're ready to lay your money on the counter and walk out with a rig.

In general it's probably best to start out with something that's not too fancy. A good but plain rig is less expensive than the fancy many-featured models, and until you've had some experience with CB operating, you're not going to know what you really want anyway. In fact, it's not at all unusual for a beginner with one of the more exotic sets to find himself all wrapped up in "cockpit trouble"—too many knobs with too many functions with too little understanding of what they do.

This is particularly true in mobile operation. Your number one obligation when you're behind the wheel of your car is to keep it from running into other cars, pedestrians, telephone poles and the like. Anything that takes your attention away from that obligation is bad, and a CB rig under your dash with a control panel only slightly

(Facing page-front.) This 23-channel transceiver has all the functions of a typical mobile in a hand-held package. An accessory auto adapter cable permits operation from an automobile's electrical system, a valuable function in high crime areas since the rig can be conveniently removed when leaving the car. (Midland International)

(Facing page-back.) RCA's Model 14T200 mobile transceiver.

Selecting A CB

When starting out to select a CB radio you first have to decide what you're going to do with it. The owner of this VW, (left) equipped with AM/FM radio and TV as well as CB, was looking at much different considerations than the driver of this big Kenworth (right) had in his choice of a rig.

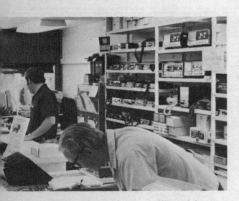

A well equipped shop like this gives you a chance to see a variety of radios and accessories plus valuable advice in making your selection.

100-milliwatt 27 MHz transceivers aren't legally classified as CB rigs though they do operate on CB frequencies. Ultra compact vest pocket models like this make very convenient short range communications tools. (JS & A Sales)

less complex than that of a jumbo jet is not conducive to safe driving. For mobile, remember the "KISS" principle—Keep It Simple, Stupid—and you'll live longer and happier, on and off the air.

The same constraints don't apply to base station operation, of course, but until you're really into CB, it's going to be tough to figure which features of a given radio are sizzle and which are steak. This is an area where *good* advice—the kind you can get from an experienced CBer, a friendly ham or a competent counterman at your local CB outlet—can save you money and even disappointment.

Hand-held CB rigs fall into two categories, licensed and unlicensed. Before you get any wrong ideas about "unlicensed," we're talking about *legal* unlicensed operation in this case. Part 15 of the FCC rules provides for the legal use of low power communications devices without a FCC license. Low power under Part 15 is one tenth of a watt or, as it's usually specified, 100 milliwatts (100/1000 of a watt). Because of their low power—1/50 of that permitted to a licensed CB transmitter—radios made under the Part 15 regulations do not have to meet the same technical requirements that licensed radios do. As a result their quality varies widely, from the $9.95 pair of kid's toys to some very sophisticated vest pocket models. One interesting legal quirk concerns these low power CB rigs which operate on 27 MHz. Because they are not licensed, it is against the rules to use one to communicate with a licensed CB station. Eventually this paradox will fade away, however. In the spring of 1976 the FCC authorized a new 49 MHz band for Part 15 communications, and 27 MHz Part 15 operation will be phased out over the next few years.

Higher power hand-held CB rigs have to meet the same technical requirements that 27 MHz base and mobile radios do. Among the licensed hand-held sets the principal differences are in transmitter power and channel capability. Hand-helds that run the maximum legal 5-watt input are quite common and generally preferred where maximum range is a critical consideration. However, their high power is a liability as well as an asset since higher power means shorter battery life and/or larger, heavier and more expensive batteries, plus a larger, heavier and more expensive radio to put them in.

The choice between a base station, mobile or hand-held radio is not as obvious as it might seem to be. Most base station rigs have provisions for operation on 12 volts for use as mobiles and from batteries during power failures or while on camping trips. Radios designed primarily for use as mobiles make fine, compact base stations with the addition of an inexpensive regulated power supply. Some CBers who have to park their cars in high crime areas or who drive a lot of rented or company cars prefer to use hand-helds for mobiles, taking the radio along when they leave the car. The decision of which type—and which make or model of that type—best serves your needs is yours. First decide what your needs really are; then decide how best to satisfy them.

The more elaborate 100-milliwatt transceivers for unlicensed (FCC Part 15) operation include all the features of their higher power brothers. (Midland International)

CONTROLS

There's one thing you should know before you can go any further on your quest for a CB rig. Just what are all those knobs and dials and what do they do? An examination of the front panel of a typical CB rig should reveal the following controls:

Volume/ON-OFF Knob: Performs the same purpose of controlling the speaker volume and turning the set on and off as it does on a broadcast receiver. The on-off knob is sometimes separate or may be combined with another control.

Squelch: The squelch circuit blanks out background noise or static when a signal is not being received, and the squelch control setting determines how strong an incoming signal must be to overcome the squelch and be heard. "Squelched" means no signal is being heard, and "unsquelched" or "open squelch" means a signal (or noise) is being received.

Channel Selector: Selects which channel the set is tuned to.

*Automatic Noise Limiter (ANL) or Noise Blanker: Reduces or eliminates noise impulses such as those

This 5-watt, 6-channel hand-held has all the features of many smaller mobile and base station transceivers. Note the HI-LO power switch to preserve batteries when conditions permit. (Royce)

Another approach to portable operation is an accessory battery pack powered mobile rig like this. (Midland International)

---

*These controls are not always provided, particularly on compact mobile rigs, but their omission does not indicate a lack of quality.

The more elaborate mobile rigs like this Horizon 29 also boast an impressive number of controls. (Standard Radio)

All those knobs aren't absolutely necessary in order to communicate. This very compact mobile rig does a fine job despite its minimum control panel. (SBE)

If CB is an occasional tool for you, then you won't need the extra functions or even 23 channels that most radios offer. For travelers this very low-cost, 3-channel model set up for channels 9 (emergency channel), 19 (highway travel channel) and another channel for rag chewing would probably prove entirely adequate. (Midland International)

generated by automobile ignitions. Of the two types the noise blanker is more effective since it actually cuts off the receiver audio output during a noise pulse. A noise *limiter circuit* merely limits the noise peaks to the same level as the received signal; therefore noise can still be heard in the background.

*Clarifier or "Delta Tune": Provides fine tuning of the receiver to "peak up" signals that are slightly off frequency. The clarifier control will only be found on SSB type CB transceivers and is much more critical to SSB operation than the Delta Tune is to AM transceivers.

*PA/CB: Switching to permit the modulator of a mobile CB transmitter to be used as a "hailer." A separate speaker mounted inside the car grille is required for PA use.

*Distant/Local: Permits reducing the sensitivity of the receiver to distant interference or overload while listening to strong local signals.

*RF Gain: Another more finely adjustable means of changing receiver sensitivity to accommodate extreme variations in signal strength.

Other controls are also sometimes provided, but those described above are the ones you're most likely to encounter, and most of them will be found on all CB radios.

One crucial factor that is all too often overlooked, particularly in the choice of a first rig, is *usability*. Are the controls on the rig you're considering clearly labeled, large enough to see and grasp and separated sufficiently so you can operate them conveniently? In a mobile rig, are the controls laid out in such a way that you can find the volume, squelch and channel selector without taking your eyes off the road? In short, does the radio have "human engineering?" If not, you don't want it. A hard-to-use radio is a constant source of irritation!

SPECIFICATIONS

When you're in the process of choosing a new rig, one of the principal things you'll need to know is how well it performs compared to other competitive radios. We're now talking about rather small but significant electrical differences that few of us have either the know-how or expensive equipment to test. Even if we did, we wouldn't have the opportunity to go through a time-consuming series of tests on each of the sets we've decided are in contention for our dollars. Therefore we're forced to fall back on this data as provided by the manufacturer in the form of his published specifications. This information is supposed to be accurate, and most of it probably is. However, it's to the manufacturer's advantage to have his radio look as good as possible, so published data is sometimes the result of measurements made on a specially selected example that's been tuned to a gnat's eyebrow by a top technician. Recent FCC tests of 25 CB radios representing a good portion of the CB manufacturers and importers found not one that completely met all of the specifications that had been supplied to the FCC!

Nonetheless, the manufacturers' claims are not all that bad and do provide a basis for comparison—the best basis we really have. Let's take a look at those specs, what they are and what they should mean to you when you're trying to make a buying decision.

There isn't much in the way of operating controls that this Cobra CAM-89 AM base station doesn't have. (Dynascan)

RECEIVER SPECIFICATIONS

Sensitivity: The measurement of the receiver's ability to pick up weak signals and reproduce them. It is measured in microvolts (millionths of a volt). The fewer the microvolts (symbol uV) the better the receiver. A sensitivity spec expressed only in microvolts is not really complete, however. It should also say how well the signal actually overcomes the background noise, expressed as the "signal-to-noise ratio" [S/N or (S+N)/N] in decibels (dB). A complete sensitivity spec should read something like ".4uV for 10dB (S+N)/N," which would mean that this particular receiver can detect a signal as weak as .4 microvolts when the signal-plus-external-noise to receiver generated noise is 10 decibels. The lower the microvolts and the greater the number of dB the better the receiver will hear. For 27 MHz .75uV with a 10dB or more signal-to-noise ratio is good enough, and even 1.5uV is adequate. (Note: Do not compare the sensitivities of two CB models unless the dB readings are the same.)

The "Formula D" CB mobile from SBE.

Royce Model 1-658 mobile rig. Note the gain control built into the microphone case.

Selectivity: Indicates how well the receiver will be able to reject signals from adjacent channels. A typical spec might read: "Selectivity: 6kHz at −6dB"—the narrower the bandpass (the channel width the receiver will receive, in kHz) and the greater the rejection (dB) the better. Another common way of specifying selectivity is "adjacent channel rejection—60dB at ± 10kHz." As before the fewer the kHz and greater the dB the less interference you'll suffer. Ideally you'll find

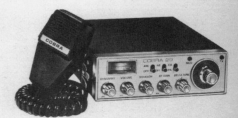
This Cobra 29 mobile rig has a front panel mike gain control. (Dynascan)

Selecting A CB

One of the features of this mobile is a built-in adjustable TVI trap. (Royal Sound)

Under-dash mounted CB mobile from Metro Sound.

This mobile has a large LED readout for channel indication and can be used with a telephone handset instead of microphone and speaker. (Boman Industries)

This very compact unit has a row of LED lamps instead of a meter for signal strength and output indication. (Fieldmaster)

This compact mobile still covers all 23 channels. (Pearce-Simpson)

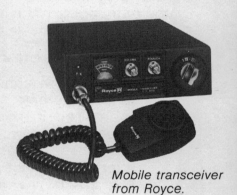

Mobile transceiver from Royce.

Motorola's Mocat Model 2000.

Another Mocat, the Model 2005, also includes a noise blanker. (Motorola)

Royce Model 1-653 mobile transceiver.

Metro Sound Model MS-1325.

Motorola Mocat Model 2020 features an LED channel indicator and noise blanker.

Cobra 21 mobile tranceiver by Dynascan.

CITIZENS BAND RADIO DIGEST 41

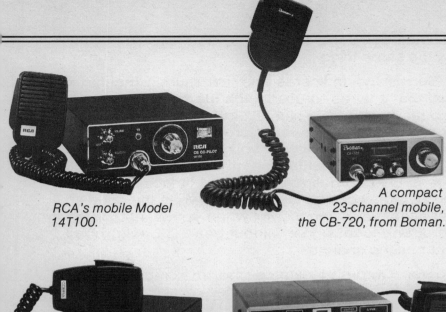

RCA's mobile Model 14T100.

A compact 23-channel mobile, the CB-720, from Boman.

Sharp's CB-700 transceiver.

Cobra 19 from Dynascan.

Metro Sound's Model HA-23C.

Boman's Model CB-535.

Royce's Model 1-660 features an LED channel indicator.

Pearce-Simpson's compact "Tomcat 23" mobile.

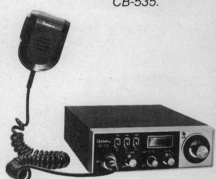

The Model CB-755 from Boman has a meter circuit to make SWR measurements.

Royce Model 1-655.

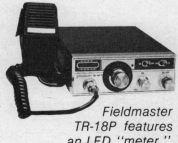

Fieldmaster TR-18P features an LED "meter."

Boman Model CB-750.

Royce's Model 1-662 features an LED channel readout with adjustable brightness and a gain control built into its mike.

Boman Model CB-740

Tenna offers three different mobiles for CB use.

Model CB-765 mobile from Boman has an LED channel indicator.

both methods used together since they are not exactly the same thing.

Squelch Sensitivity: Indicates the minimum amount of signal that it takes to unsquelch the receiver so you can hear an incoming signal. Squelch sensitivity must be greater (that is, fewer microvolts) than the rated sensitivity of the receiver or you won't be able to hear a weak signal without turning off the squelch. *Squelch Range* is also sometimes specified. It indicates how strong a signal can be blocked out when the squelch is set at its maximum, useful when you want to listen for a strong local signal without hearing background chatter on the channel.

Audio Power Output: Indicates the maximum power available to drive the radio's speaker. Included with the audio power rating is (usually) the maximum distortion. The greater the power and the lower the distortion the better, though remember CB is not hi-fi and 2-3 watts at 10 percent THD (total harmonic distortion) is usually adequate unless your set has a tiny speaker, badly mounted, and you're using it in a noisy truck cab.

Automatic Gain Control (AGC): A circuit used in a receiver to keep the audio output to the speaker relatively constant despite changes in the level of the received signal. AGC performance is usually expressed as "Output varies no more than 6dB when input changes from 1uV to 100,000uV." The less the output changes in dB the better that receiver's AGC performs.

You may find additional receiver specs provided—for example, "image rejection," the ability to reject strong signals from outside the CB band expressed in dB (the higher the better) is one. If you're looking at a hand-held unit or plan to operate a mobile rig for extended periods without the engine running, the lower the *current drain* with receiver squelched and unsquelched the better.

TRANSMITTER SPECIFICATIONS

Power Output: The radio frequency energy the transmitter can deliver. FCC rules limit a Class D CB/AM transmitter to 5 watts maximum DC power supply input to the final amplifier stage and 4 watts RF power output from that final amplifier. Note that the 5 watts is *power supply* input to the final amplifier, and since a certain amount of energy is lost—usually as heat—in a typical power amplifier and 80 percent efficiency is about as high as technically practical, 5 watts DC input means about 4 watts maximum RF output. For a good CB mobile or base station 3.5 to 4 watts is normal. A listener could never tell the difference between the two.

Spurious Output: Is the amount of signal output the transmitter has on frequencies other than the CB channel you are transmitting on. FCC rules for 23-channel radios require spurious outputs be at least 49dB lower than the carrier power, and the new 40-channel sets must have spurious outputs down at least 60dB. The higher the number, the lower those unwanted signals are—and those signals are the ones that chew up the neighbors' TV!

Modulation: Is the amount of modulation (voice) the transmitter is capable of applying to the carrier. The ideal would be 100 percent, but less than that is preferable since exceeding 100 percent ("over modulating") causes a rough, broad signal. Ninety percent modulation is typical. A statement like "Internal limiter circuit prevents exceeding 100 percent modulation" indicates the presence of a special circuit that prevents over modulation.

The channel indicator and meter in this Midland 13-862 are mounted at an angle for easier viewing.

Boman Model CB-725:

GENERAL SPECIFICATIONS

In addition to the electrical specifications discussed above, there are some other specs you're going to want to compare. Size and shape are often important in a base station installation, but they are crucial in an automobile. Many of today's cars have little if any room under the dashboard for another radio. You could very well find your choice of sets restricted to the few capable of fitting into the space you have available. A mobile rig always comes complete with a mounting bracket. Is the bracket supplied compatible with your installation needs or desires? Of course there are other alternatives, for example accessory mounts that go on the floor or hump.

Is the internal speaker large enough and mounted in such a way that you're going to be able to hear it? Even with the windows open and the speedometer at 55 mph? If not, you're going to have to add an outboard speaker. Is there a meter (or meters) on the front panel? If so, it undoubtedly reads received signal level on receive, but on transmit may read only battery voltage instead of sampling the transmitter output. Transmitter RF output sampling is a much more valuable indication since it will change if either

Cobra 138 from Dynascan is an SSB as well as AM transceiver.

The Model 13-893 is a mobile that can operate both AM and SSB. (Midland International)

44 Selecting A CB

Boman Model CB-770 is a mobile capable of SSB and AM operation.

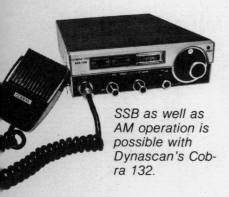

SSB as well as AM operation is possible with Dynascan's Cobra 132.

The Cobra 139 is an SSB/AM base station. (Dynascan)

Pearce-Simpson's "Simba" is an AM/SSB rig with digital clock.

the antenna tuning or the transmitter output change and can thus help identify and diagnose problems.

One last consideration. Though not exactly a "specification," you should consider what the manufacturer's warranty policy is and how available his service is. A radio that can be repaired locally with someone else footing most or all of the bill is always a better buy than an "orphan."

SSB vs AM

Even if you don't run into it much before you get on the air, you're going to hear a lot about single sideband (SSB) once you're an active CBer. In several respects SSB is the way to go, but in many ways it is not. Before getting into the whys, however, let's take a look at what SSB is. To do that we'll also have to review what AM is.

When the power of a radio frequency signal (the carrier) is varied at an audio frequency rate, the result is the generation of two sidebands, one on either side of the carrier. This process is called amplitude modulation (AM). Since the intelligence in the signal is contained in either sideband, one sideband can be eliminated without affecting communications. The carrier, though useful in detecting (recovering the modulation) the signal at the receiver, can be supplied by the receiver so it too can be eliminated at the transmitter. The result is single sideband suppressed carrier or SSB for short. It takes up only half as much spectrum and—since all the power is concentrated in one sideband—is much more efficient.

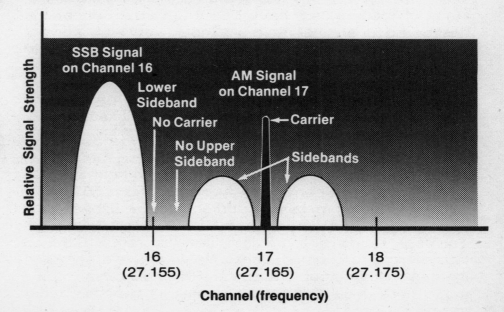

Why then, you might ask, doesn't everyone use SSB and eliminate AM? There are several very good reasons for this. An SSB rig is considerably more complex and thus more expensive than an AM rig, and an SSB signal cannot be received understandably on an AM receiver. SSB is sometimes nicknamed "Donald Duck" simply because it sounds like Walt Disney's famous quacker when received on an AM receiver. SSB signals must be tuned in very carefully to be understood, and even with the best of tuning a voice heard over SSB does not sound as natural as the same voice going over AM.

The Cobra 135 SSB/AM base station with digital clock. (Dynascan)

The typical CB/SSB radio is actually a combination transceiver, capable of operating on either conventional AM or SSB and using either upper or lower sideband with the flip of a switch. The advantages of SSB are realized primarily when everyone else using the same channel is also using SSB—otherwise AM carriers can severely restrict SSB range by causing annoying squeals (heterodynes) in the SSB receiver's detector circuit. Channel 16 is widely used as an SSB channel, and most AM users try to stick to the other 22 channels. Since SSB uses only half of a channel, SSBers do get double mileage out of their rigs—different contacts going on simultaneously over the same channel, one on upper sideband and the other on lower.

The telephone handset with this Midland Model 13-886 AM base station provides privacy for the operator and minimizes disturbance to the rest of the family.

SSB is also more efficient since all of the transmitter power (12 watts peak allowed by the rules) is in the part of the signal that carries the voice. As a result, the maximum range of SSB (under those rarely realized "ideal" conditions) is considerably greater than AM can achieve. Under practical conditions, the benefits of SSB are not so easy to achieve.

The "Bearcat 23C" from Pearce-Simpson features a digital alarm clock.

The plain facts are that most CBers are looking for simple, easy-to-use radios that do the job at minimum cost. AM fits that description, but SSB does not. By far the majority of CB/SSBers are CB hobbyists, enjoying radio for radio's sake like amateur radio operators. Most CB communications is on AM and will continue to be for a long time to come.

It's quite possible that someday the FCC will make Class D CB part or even all SSB because, if for no other reason, such a move would double the number of channels available. However, don't expect such a move to come soon if it ever comes. With tens of millions of AM/CB rigs in use you can expect they'll be usable for many years before they will be put off the air by a rule change.

Royce's AM base station has a controllable intensity LED channel indicator and separate meters for each function.

Selecting A CB

A compact AM base station is the Cobra 85 from Dynascan.

Pearce-Simpson's "Super-Lynx" includes a built-in digital alarm clock.

AM and SSB operation and digital clock are all features of Midland's Model 13-898B.

In-dash mounted CBs both look better and are harder to rip off. This Boman CBR-9300 includes an AM broadcast radio.

Tenna's A1301 is an in-dash combination 23-channel CB and AM/FM stereo receiver.

AM/FM broadcast reception complements Boman's CBR-9600 in-dash CB set.

Clarion's Model RE-366E provides both AM and FM broadcast reception when the in-dash CB is not in use.

J.I.L.'s in-dash Model 852CB is a complete "entertainment center" with stereo tape deck as well as AM/FM broadcast.

The LED digital readout on the Boman Model CBRT-8800 gives digital frequency readout on AM and FM as well as channel indication on CB. It also has a stereo tape deck.

The Model CBR-9500 from Boman has both AM and FM broadcast reception, and an optional telephone handset is available for greater privacy when using it for CB transmission.

AM and FM multiplex plus stereo tape deck complements the J.I.L. Model 606CB transceiver.

NEW CHANNELS AND "OLD" RADIOS

The long awaited expansion of the 27 MHz Class D CB band from the original 23 channels to 40 channels was announced in July, 1976, by the FCC. The expansion actually goes into effect on January 1, 1977, and until then no 40-channel radios may be sold. Fifteen of the 17 new channels are higher in frequency than the original 23 with the highest being 27.405 MHz. The other two are sandwiched in between the existing channels 22 and 23.

This channel expansion seems to put the CB newcomer or even the active CBer who wants to upgrade his equipment into a real quandary: Why buy a new piece of equipment when it's going to be obsolete in a short time?

The answer is that a 23-channel CB rig *won't* become obsolete on January 1, 1977! With perhaps 20 million 23-channel CB rigs in use, a CBer with "only" 23 channels to talk on is not going to hurt for people to talk to. Channels 1 through 23 are where the action is today, and they are where the action will continue to be after the 17 new channels become available. Even after new 40-channel rigs are on the market, it's going to be some time before any appreciable quantity come into use, and in the meantime there are several good reasons why a 23-channel set will be your best buy.

The new 40-channel sets will be more expensive, probably a good deal more expensive. Not only will they cost more because of the expense of their increased coverage, but as part of the channel expansion package, the FCC has required the new rigs to meet much tougher technical requirements for both transmitter and receiver. Designing, building and testing radios to tight technical specifications costs money, and in the end it's you, the buyer, who foots the bill.

Another factor working in favor of the 23-channel radio buyer is dealer and manufacturer worry that their 23-channel stock has suddenly become obsolete. The result has been some remarkable price cutting, and the CBer who wants to shop around can find some super bargains in first rate deluxe grade 23-channel rigs. And after the 40-channel radios are actually on dealers' shelves, those deals will get even better. Though you'll eventually want to move up into those inviting new channels, the 23-channel radio you buy today and eventually retire to the second car, boat or simply "back-up" service sometime in the future can still be the best deal you'll ever make on CB equipment!

The Clarion Model PE-670E offers AM/FM broadcast and stereo tape listening with its in-dash mounted CB.

Though it looks fairly conventional, all the primary operating controls including volume, squelch, channel selection and LED readout are mounted on the microphone of SBE's Model 32 CB.

Midland's remote control Model 13-955 divides its operating controls between the microphone and its compact control head.

Selecting A CB

Royce's in-dash CB is also an AM/FM broadcast receiver.

Royce's remote mount radio has channel selection and indication on the microphone, other controls on the control head.

"Wait a minute!" you may say, "Why can't I buy a 'converter' or have some technical whiz convert my 23-channel radio to work on the new channels?" Because the FCC has expressly forbidden such actions, that's why! They've got a good reason . . . you've certainly heard of the growing problem of CB (and other) radio transmitters interferring with TV, FM broadcast and even police, fire and aircraft radios. It is that very problem that caused the FCC to require makers to tighten up their specifications, and the commission is not about to permit older, interference-generating radios to move over into the new channels where their harmonics (multiples of the operating frequency) fall into previously unbothered TV, FM and two-way radio channels.

There are a couple of exceptions to the prohibition against conversion, however, and they're worth knowing about. The first is that manufacturers will be able to market 40-channel radios that meet the new technical specs before January 1, 1977, providing that they are internally set up so that they work on only channels 1 through 23. Then after January 1, 1977, the buyer will be permitted to return the set to the manufacturer for expansion to full 40-channel coverage *and* retesting to make certain that it still meets specs. Though no one has worked out all the details on how this will be carried out as this book goes to press, several manufacturers have stated that they feel a $25 to $50 service charge (in addition to the purchase price) would be required.

The FCC has also given manufacturers another alternative for their 23-channel transceivers—take the older model radios in as trade-ins and rebuild them into "new" 40-channel radios that meet the tough new specs. Though many current models are not suitable for such upgrading, some are. Which ones are suitable remains to be seen. If the 23-channel radio you have now or choose to buy is one that can be upgraded, you're lucky. It'll certainly be a while before anyone knows for sure which ones can . . . but it can't hurt to ask when you're out shopping.

Summing up, the CB picture is changing and will continue to change, but the action is *now*! You can wait for the ultimate . . . but just remember that the old-timer waiting for the horseless carriage to be perfected is still riding a horse!

THE CHANGING FACES OF CB MOBILES

Until quite recently CB mobile rigs didn't look all that much different. They had a channel selector, a few other knobs, perhaps a meter and a light or two, all on a front panel designed to stick out from under a car dashboard. Most rigs still are that way, but now there are a growing number that aren't.

Probably the first of the new generation radios was the "in-dash" model. In-dash CB sets have two reasons for existence, security and appearance. Getting a radio out from under a dashboard is usually a fairly simple job, particularly for an experienced thief. Getting one *out* of a dashboard is an entirely different matter.

An add-on accessory under a dashboard always looks like what it is, added on. Wives in particular cast a jaundiced eye on such items, whereas they seem to feel much more agreeable to Dad's toys if those toys look like they belong. In-dash mounted CB rigs fit that bill, and just so you won't have to forego the pleasures of broadcast, FM or even stereo tapes, in-dash CB packages including those functions are now available.

Moving the operator functions away from the rig proper and putting them in a more convenient spot is an even more recent innovation. Several makers now have made the microphone into a control head, complete with volume and squelch controls and even channel selection. Security is one reason for this innovative idea. It's a lot easier to hide a microphone than it is a complete radio after you've parked your car, and the radio itself has already been mounted up behind the dash, under the seat, in the trunk, or in some other "invisible" spot. However safety and convenience are equally important reasons why the control head-microphone combination is catching on the way it is. Being able to bring the controls up in front of your face instead of peering or groping under the dashboard is a lot safer when you're trying to maneuver through heavy traffic.

Because the would-be CBer has so many options open to him when it comes to choosing a new rig, it pays to do some research before you reach for your wallet. Get literature from a local CB shop or write to the various manufacturers—you will find a pretty complete listing of them, complete with addresses, in the Manufacturer's Directory in the back of the book. Do your homework and you won't be unhappy with the results!

chapter 5

ANTENNAS FOR CITIZENS BAND

you're going to put that on our roof?

Your antenna is the most important part of your CB unit's installation. With a good antenna properly installed, you'll be able to communicate effectively using even the least expensive, low end radio. A poor or improperly installed antenna, on the other hand, won't give good results with any radio at any price. It can even damage the transmitter.

Before getting into specifics, let's talk about a little theory. We've all heard that light waves travel at a speed of 186,000 miles per second. Radio waves do too. 186,000 miles or 300,000 kilometers or 300,000,000 meters is the distance that our radio wave travels in one second. The frequency of the radio wave from our CB transmitter is approximately 27 MHz or 27,000,000 cycles—alternations from plus to minus and back again—per second. If we divide the distance the wave travels in a second (300,000,000 meters), by the number of cycles the radio wave goes through in a second (27 million), we come up with the number 11. That's the *wavelength* of a CB signal—*11 meters*—and where that other name for CB came from.

$$\text{Wavelength} = \frac{\text{speed of light}}{\text{frequency}} = \frac{300,000,000 \text{ meters per second}}{27,000,000 \text{ cycles per second}} = 11 \text{ meters (about 36 feet)}$$

Whether a mobile CBer plans to operate from a little 4-wheeler around town or from an 18-wheeler on cross-country jaunts, he still needs to pay attention to antenna basics.

In order to radiate a signal most effectively an antenna must be a *half wave* or a multiple of a half wave. For the CB band, then, the basic antenna is 5.5 meters or about 18 feet long. Making the antenna precisely a half wave long permits the radio wave to alternate from plus to minus to plus and so on without part of the signal being reflected back down the feedline to the transmitter. Now you know what's going on when you adjust an antenna for *minimum reflected power* using a VSWR meter. You're tuning it so its electrical length is precisely a half wave. Half wave? But the box said "quarter wave" on it! Aha, time for a little more theory.

In order to operate as an effective, efficient antenna a half wave antenna doesn't really have to be a nicely balanced, symmetrical structure. In fact the earth (ground) or any large electrically conducting surface (ground plane) can replace one half of our half wave antenna very nicely. The body of a car does a pretty decent job as a ground or ground plane, replacing half of our half wave antenna so all we now need on the back bumper of our mobile is the other half, a 9-foot whip. A similar antenna, the "ground plane," is also sometimes used for a base station antenna. It uses several quarter wavelength long *radials* to replace the car body as its ground plane.

Now that we know what an antenna is, let's briefly consider some of the problems that we're going to encounter in our attempts to select a good 11 meter antenna and get it working. The first problem is going to be selecting the antenna itself. Whether you're shopping for a mobile or a base station antenna, beware! Antenna specifications are the technical area in which the most nonsense is tossed around, but there is really no "black magic" about antennas. A CB mobile whip that's 9 feet long is a more efficient antenna than one that has been shortened physically by the use of a loading coil, and a full-sized half wave antenna from one manufacturer should be just as effective as a full-sized half wave from another, as long as the antennas are properly constructed. One recently released data sheet claimed "16dB gain" for its half wave antenna. Since each 3dB gain is the same as doubling the transmitter power output, the antenna maker was trying to give the impression that his antenna would give the effect of boosting your transmitter power by a factor of 40. That's nonsense!

What wasn't said in that data sheet was what he used for comparison, since to get "16dB gain" his reference antenna must have been a damp noodle or even a section of short circuited feedline. Most manufacturers will state their antenna gain specs as "6.7dB gain referenced to a half wave dipole" or "gain 9dB over isotropic" (an "isotropic" antenna is a theoretical antenna that is about 2dB worse than a half wave dipole. When you see an isotropic reference used in specifications, simply subtract 2dB and you'll be able to compare performance with another antenna that was measured against a half wave).

Once you get past the "specsmanship" of performance data, what basis should you use for selecting an antenna for your new CB? Here are some good starters:

- Quality materials that won't rust, corrode or rot
- Sturdy construction, capable of surviving wind and ice
- Ease of installation
- Attractive appearance
- Accessibility for service or adjustment
- Manufacturer's policy for replacing defective parts
- Manufacturer's overall reputation

These criteria apply pretty well to any antenna, base or mobile, so keep them in mind when you visit your friendly local CB store. Don't overlook the advice a sharp counterman can give you, either. Just remember he makes more money from more expensive merchandise, but a bigger price tag doesn't necessarily mean a bigger signal. Friends already well-established in CB can also be helpful in your antenna search, as can friends who are hams.

Finding a good location for your CB antenna is another problem you're going to have to consider. Proper antenna location is just as important as the antenna itself. The basic rule is "the higher the better." However, there are some other factors that need to be considered such as the FCC limits on the height of base station antennas. Nearby objects can detune an antenna and cause "dead spots" in its pattern. Trees absorb radio frequency energy and will cut down both received and transmitted signals. Nearby power lines, electrical machinery, cars on a busy street and even TV sets (or their antennas) can put enough noise signal into a CB receiver to wipe out all but the strongest signals.

Feedlines can be another problem area. Even the best quality coaxial cable absorbs some of the signal passing through it (called "attenuation"). The thinner cables, though cheaper to buy and easier to use, are much worse than the heavier cables. Cable *quality* varies, too—some amazingly bad coax is being made for the CB market.

> "Finding a good location for your CB antenna is another problem you're going to have to consider. Proper antenna location is just as important as the antenna itself."

"Snap-in" mount mobile antennas can be used on rooftops where they'll provide a nearly uniform pattern in all directions or rear decks or trunk lids where they'll do almost as well. This example features a high efficiency, extra large loading coil. (Antenna Specialists)

For hatchbacks a "no-hole" mounting bracket that can be adjusted to make the antenna vertical is available. (Antenna Specialists)

A bumper mount like this one is the best place for a full-size 108-inch whip.

MOBILE ANTENNAS

So much for the general antenna considerations. You've got a license, and you've got a rig. Now you want to put an antenna on your car so you can use them. A mobile installation is going to be a compromise at best—you're going to want to be able to drive your car as well as talk from it!

The ideal mobile antenna from an electrical point of view is a quarter wave mounted in the center of the car roof. It is the most efficient and provides a uniform pattern in all directions ("omnidirectional"). As a practical matter such an antenna is highly impractical if not impossible since city streets, gas station marquees and parking garages are simply not designed to provide 13-plus-foot clearance for a CBer's antenna! However, such installations are not unknown in the open spaces of the West and Southwest where ranchers, hunters, prospectors and campers need all the range they can get for their communications.

For the rest of us the choice of a mobile antenna is limited by the practical consideration of maximum height. Three possible locations, rooftop, trunk lid (or rear deck) and bumper are commonly used, but the higher you go on the car body the shorter (and therefore less efficient) your antenna is going to have to be. A full length whip is still awfully high—10 feet or more from the ground—when bumper mounted, so a number of CBers who use them keep them bent over the top of the car and fastened to an insulated clip on the rain gutter most of the time. Such an arrangement, though inefficient, is still good enough for short range communications around town, and it takes only a moment to release the antenna for normal use when going on the road.

BUMPER MOUNTING

Bumper mounting is often used for shorter antennas as well as for full-size 9-foot whips. A 4- or 6-foot loaded antenna is more efficient than even shorter radiators, yet permits

Two types of bumper mounts; the longer for full-sized cars and the shorter for compacts. (Avanti)

very low clearance when bumper mounted. Bumper mounting does have, however, one significant disadvantage—it distorts the radiation pattern so the signal strength is not uniform in all directions. The maximum signal occurs in front of the car toward the corner opposite the one on which the antenna is mounted, while the weakest signal goes toward the rear. Despite its distorted pattern, however, a bumper mounted antenna is still a good choice for mobile use.

ROOFTOP MOUNTING

Mounting the antenna in the center of the car roof provides the most uniform signal in all directions but also usually requires the shortest and least efficient antenna. The car roof is also invariably the most difficult place to mount the antenna permanently since a hole has to be drilled through the roof and the antenna lead-in fished through the head liner. One advantage of a rooftop mounted antenna is that it's less accessible to vandals and thieves than an antenna anywhere else on the car. One disadvantage that tends to be exaggerated is the effect of the hole on the resale value of the car. Snap-in plugs are available that can be painted to match the car and do quite a satisfactory job of filling in the hole. In the case of a luxury car where a more permanent repair is preferred, it's still not a big project, typically $20 for patch and repaint.

TEMPORARY MOBILE ANTENNA MOUNTS

For those who prefer rooftop antennas but who may use a number of different cars or simply don't want to drill a hole, there's still another option, the magnet mount antenna. Though a magnet mount antenna is not generally quite as efficient as the same antenna permanently mounted, it is a good alternative. Having the lead-in come through the window is unsightly and often inconvenient, but by the same token the antenna can be easily and quickly removed and stored in the car when it's parked, preventing the antenna from being stolen and also concealing the fact that the car

A rooftop installation is the neatest appearing, most uniform in radiation and toughest to install of all mobile antennas.

Magnet mount antennas, usually thought of as a temporary expedient, are often used semi-permanently by CBers who prefer rooftop mounting but who often go into low clearance garages. They're also easily removed for floor or trunk stowing in high crime areas.

A variety of heavy mounts are available for trucks and other rough duty vehicles. (Avanti)

Clamp-on gutter clip mount with center-load antenna. (Antenna Specialists)

Detail of semi-permanent gutter mounts. (Antenna Specialists)

This camper driver has a secondary advantage from his mirror-mounted twin antennas—the brightly colored balls remind him where he parked in parking lots and shopping centers.

is CB equipped. With a reasonable base- or center-loaded antenna and today's highway speeds, there's practically no chance of a magnet mount antenna blowing off, either.

Another type of mobile antenna mount that doesn't require chopping a hole in the car is the gutter clip mount. These are made in both semi-permanent (screw-fastened) and spring-loaded quick remove versions. Gutter clip mounts are definitely a compromise, for though they put the antenna up high where it should be, their position on the side of the car roof produces a distorted radiation pattern. Gutter clip mounted antennas are also often seen mounted in pairs on opposite sides of the car roof, an impressive looking setup like the truckers use. Unfortunately, even when fed by a properly adjusted phasing cable harness, phased antennas on an automobile roof are too close together to be very directional—and what fore and aft directivity they do have is not actual gain but only loss of signal to the sides.

REAR DECK MOUNTING

Probably the most popular spot to mount a CB antenna on a car is somewhere on the rear deck. Rear deck mounting has a lot of advantages not the least of which is that it is low enough on the car that the antenna can be long enough to be reasonably efficient, yet it's high enough that the antenna radiation pattern is pretty uniform in all directions. It's also an easy place to work with ready access to the inside of the mount for installation and cable routing. If you don't want to chop a hole in that pretty sheet metal, the rear deck is the place to go. Several types of trunk lip or trunk rain groove mountings are available that work just about as well as the more permanent mounts but don't scar the car. One caution with these mounts, though—it's awfully easy to pinch the coax lead-in with one of these mounts and that's nothing but trouble.

Trunk lip mounts provide a neat installation with feedline inside the car without the necessity for cutting any mounting holes. (Antenna Specialists)

This trunk groove mount can also be used for antenna mounting on hatchbacks or at the rear edge of the hood. (Anixter-Mark)

Heavy duty trunk lip mount from Anixter-Mark.

Mirror mounts are generally used only on trucks, campers or other larger vehicles. (Anixter-Mark)

A disguise antenna has no loading coil so looks exactly like the BC antenna in whose mounting hole it fits. This helps avoid tipping rip-off artists that the car is CB-equipped. A special coupler lets it operate on both CB and the broadcast band. (Antenna Specialists)

CONCEALED ANTENNAS

Concealed antennas are becoming more and more popular with CB operators as the CB rip-off problem escalates. These take several forms, the most elaborate being a motor driven telescoping antenna that can be mounted inside the rear fender or even (in some cars) the front fender. When not in use, the antenna is usually fully retracted. It attracts little attention and looks pretty much like a regular car broadcast antenna. With the erecting motor wired to the auto accessory switch or the CB rig's off-on switch, however, it automatically extends ready for use whenever needed.

Several manufacturers make CB antennas that look just like broadcast antennas and so do not tip off would-be thieves that the car is CB-equipped. Antenna tuners which can be adjusted to make the regular car broadcast antenna work as a reasonably efficient 27 MHz antenna are also available, as are trunk edge mounts that permit the antenna to be folded into the trunk and out of sight.

So far we've talked mostly about mounts and very little about the antennas themselves. What makes a good antenna, and how can you tell a good antenna when you see one?

This motor-driven telescoping CB antenna doesn't look like a CB antenna and also foils vandals. (Tenna)

A front fender mounted CB antenna can sometimes utilize the hole drilled for the original broadcast antenna. The CB antenna is then used for broadcast reception as well as CB, either through a coupler or with a switch. (Antenna Specialists)

This pickup truck's owner worked out his own mount and nicely avoided cutting or drilling the vehicle's sheet metal.

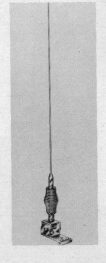

Bumper mount with heavy duty spring and 108-inch stainless steel whip. (Antenna, Inc.)

MOBILE ANTENNA PATTERNS

Where you mount the antenna on your car or truck *does* make a difference in where the signal goes and doesn't go. However, a long, efficient antenna will often work better in its worst direction than a short inefficient one will in any direction.

Putting your antenna at the center of your car roof provides the most omnidirectional coverage.

Trunk lip or rear deck mounting is also good, though it favors the front of the car.

Moving a rear deck antenna to the side or onto the fender delivers the most signal to the direction of the opposite front fender.

Mounting the antenna on the bumper produces a pattern even more directive toward the front and side.

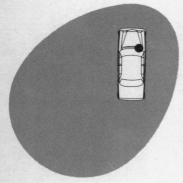

Cowl mounting puts the maximum signal to the rear and over the opposite fender.

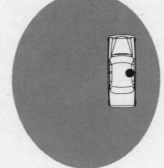

Mounting the antenna on the rain gutter gives about the same signal fore and aft, though it's strongest to the opposite side of the car.

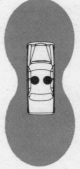

Dual antennas fed with a phasing harness like the truckers use extends the signal fore and aft with little to the side. If you can't get the antennas at least 7-8 feet apart, however, don't bother!

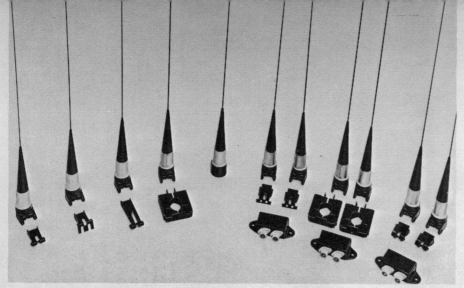

Most antenna manufacturers offer a variety of antennas and antenna mounts. This is Motorola's "Mocat" line.

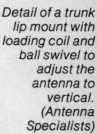

Detail of a trunk lip mount with loading coil and ball swivel to adjust the antenna to vertical. (Antenna Specialists)

CHOOSING YOUR MOBILE ANTENNA

Advertising claims notwithstanding, there is no substitute for length in predicting how well an antenna is going to perform. That unwieldy (and ugly, to some eyes) 108-inch steel or fiberglass whip will outperform any shorter 27 MHz mobile antenna because anything less than 108 inches is a compromise. This isn't to say that an antenna that's shortened by means of a loading coil can't be a pretty darn good antenna, though. You just have to realize that shortening your antenna (and how you shorten it) *will* reduce performance compared to the full-size version.

Though an antenna can be physically shorter than a quarter wavelength (our 11 meter 9-footer) it must still be a quarter wave electrically if it is to work efficiently or even at all. This is accomplished by *loading* the antenna, a tuning process that extends its electrical length. For mobile antennas this is accomplished by either *linear loading* or *loading coils*. Actually the two are not really all that different... a linearly-loaded antenna is its own loading coil wound the full length of the antenna. The antenna wire spirals around a rod (usually fiberglass), and the whole assembly is then covered by a vinyl sheath. The result is a rugged though not too graceful antenna with good electrical characteristics.

Coil-loaded antennas are made in three forms: base-, center- and top-loaded. Since the maximum signal is radiated by the bottom portion of an antenna and a coil is a poor radiator, the base-loaded antenna is the least efficient of the three. Conversely, the top-loaded antenna is the most efficient. Unfortunately, the electrical and physical characteristics are in direct conflict—the antenna that works the best is least attractive and is most susceptible to

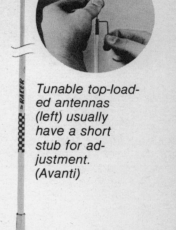

Tunable top-loaded antennas (left) usually have a short stub for adjustment. (Avanti)

Helically wound "Heliwhip" antenna from Anixter-Mark.

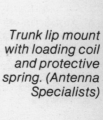

Trunk lip mount with loading coil and protective spring. (Antenna Specialists)

Center-loaded mobile antenna (right) with mirror mounting bracket.

A dual antenna (above) for truck mirror mounting. The special phasing harness provides maximum signal toward the front and the rear of the vehicle on which they're mounted. (Avanti)

Trunk lip base-loaded and gutter mount center-loaded (left and below) antennas from RCA.

Five-eighths omnidirectional antenna with quarter wave radials. (Avanti)

damage and vice versa. Compromise? The center-loaded antenna!

Of course, the base-loaded mobile whip does work and, in fact, works pretty well. The tremendous number of them that you see on cars, trucks and RVs is testimony to that. The fact remains, however, that a loaded antenna of a given length performs better as the loading coil moves higher on the antenna. What you have to consider when you go to buy your antenna is the attractiveness of the antenna, the length you're willing to mount and where you want to mount it on your vehicle!

BASE STATION ANTENNAS

Base station antennas fall into two basic categories, omnidirectional and beam. Omnidirectional antennas are just that—they radiate (and receive) signals equally well in all directions. For most CBers as well as for other two-way radio users they make the most sense since they are the least expensive, the easiest to install and maintain, and since they respond equally well to signals in all directions.

Directional or beam antennas can be very useful for several reasons. In the first place they effectively amplify both the transmitted and received signals, typically by four or more times compared to the same signal on a omnidirectional antenna. Even more important in some situations is their rejection of unwanted signals. Turning a beam away from an unwanted signal can reduce it to $\frac{1}{100}$ of the strength it would otherwise have. On the negative side beams are more expensive and more difficult and expensive to put up and keep up (typically you'll need a tower and rotator for your beam) than an omnidirectional antenna.

In most cases the decision between an omnidirectional and beam antenna will be an easy one to make since it'll be based on your own specific needs. However, it is not at all unusual to find an active CBer who uses both. Let's take a closer look at each of them.

OMNIDIRECTIONAL ANTENNAS

Many CB operators choose the 9-foot quarter wave ground plane vertical antenna for their base stations. It works well, is inexpensive in commercial versions and even cheaper if you are handy enough to build your own out of aluminum tubing or electrical conduit. If you have a metal roof, you don't even need the radials—the roof makes a fine ground plane. If you've got a high tree available, there's an even simpler, cheaper and more effective home brew antenna—the half wave wire vertical. Simply connect an insulator between two 9-foot long pieces of No. 12 or No.14 copper wire, solder the center conductor of your coax feedline to the wire on one side of the insulator and the copper braid shield to the other wire, and then hang the whole assembly from a tree branch using another insulator fastened to the other end of either of the 9-foot wires. It works! Try to run the coax feedline at a right angle to the antenna, however, because if it droops down too close to the bottom half, the antenna won't work as well.

Most commercial omnidirectional base station antennas are basically very similar to the quick and dirty home brew antenna described above. The significant difference is that they are mounted and fed from their bottom end. Some have fairly elaborate structures with various types of appendages hanging from them. Most of these are there because they look impressive or offer some mechanical benefit—they do not however add any significant boost to the signal!

What you really want to look for in an antenna for your base station is *quality*! Is the antenna made of materials rugged enough to take year 'round exposure to your weather (wind, rain, ice, salt, fog . . .)? Is it designed well, so continued wind buffeting won't loosen up telescoping parts or crack insulators? Look at some nearby chimneys —how many TV antennas have dangling or missing elements? If your CB antenna looks like that a year from now, it won't be working too well, and if it should decide to collapse on the neighbor's house (or kids), you'll really have a problem.

Once you've selected your antenna, you've still got to mount it, a job that requires answering two basic questions—where and how. Theoretically the first is easy—as high and as in the clear as possible. Practically this means for the average householder some sort of TV mount such as chimney strap or a wall mount at the peak of the roof. These are good choices if you keep one thing in mind—a 20-foot long vertical antenna in a high wind exerts a lot more leverage on its mount than the average TV antenna 3 or 4 feet above its mount. Keep your chimney straps or wall brackets 4 or 5 feet apart, and your antenna will stay up a lot longer.

BEAM ANTENNAS
For the CB operator who is willing to go to the added trouble and expense of putting one up, the beam antenna offers two significant advantages. First, it increases its effec-

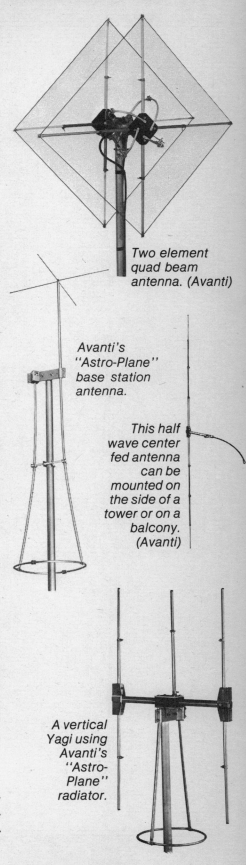

Two element quad beam antenna. (Avanti)

Avanti's "Astro-Plane" base station antenna.

This half wave center fed antenna can be mounted on the side of a tower or on a balcony. (Avanti)

A vertical Yagi using Avanti's "Astro-Plane" radiator.

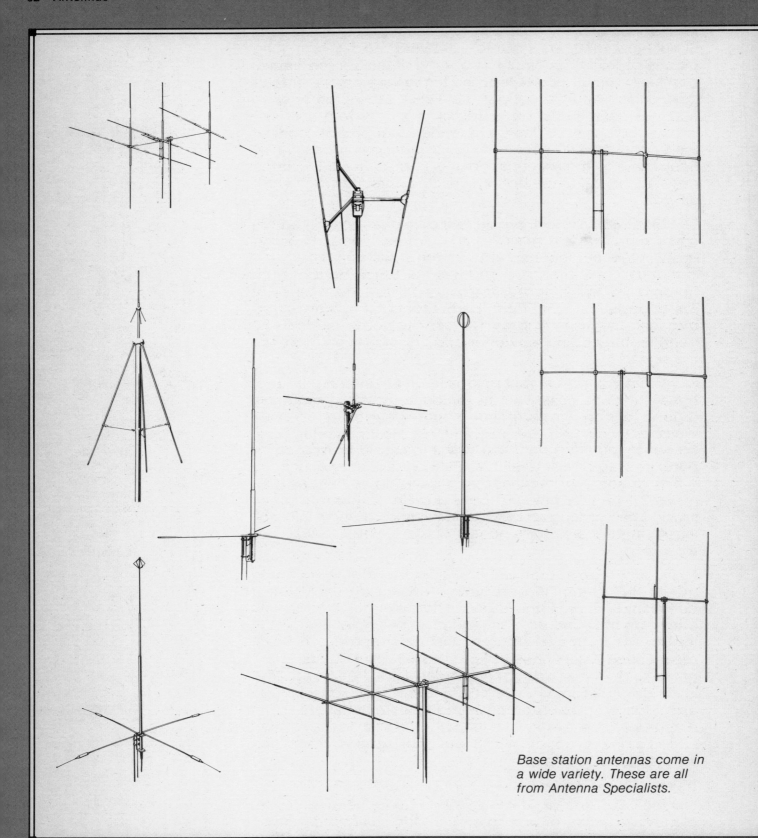

Base station antennas come in a wide variety. These are all from Antenna Specialists.

tive range on both receive and transmit, probably 25 to 50 percent because of its gain ("amplification effect") in the forward direction. Second, because it reduces interference from other directions, a beam can be even more effective in increasing usable range in heavily populated areas.

Two types of beam are used for the CB service, the Yagi and the quad. A 27 MHz Yagi consists of a half wave vertical radiator and one or more parasitic elements spaced so that they effectively direct or reflect the signal in the desired direction. The quad has its radiating element, two half waves, folded into a square which is how it came to be named the "Quad." A quad also uses parasitic elements to increase its gain and directivity.

Radio waves are polarized—that is, a receiver with a vertical antenna will receive more signal from a vertical transmitting antenna than it will from a horizontal antenna. Almost all CBers use vertically polarized antennas since verticals are much more practical for mobile use than horizontal antennas, but some operators—mostly SSB users interested in base-to-base rather than base-to-mobile communications—use horizontally polarized beams to reduce interference from vertically polarized AM stations. With those few exceptions, CB Yagis are vertical antennas and communicate best with other vertical antennas.

Because of its design the quad antenna radiates and receives a combination of vertically and horizontally polarized signals. This is not a disadvantage when communicating with a vertically polarized station, however. By the time a radio wave bounces off buildings and power lines and is filtered through trees and brush, it's no longer only vertical but a combination of polarizations. In this situation a quad often has a marked advantage, particularly with mobile signals, since it responds to *all* the signal components.

Though most beam users have rotators to point their antennas in different directions, a fixed beam can also be very useful. A housewife whose CBing consists mainly of progress reports from her homeward bound husband would do well to use a fixed beam pointed toward his line of travel, and a farmer whose farmhouse is at one end of his property could increase his own signals and reduce interference from (and to) his neighbors with such a setup.

However, most CBers have more varying needs than those and thus require rotators to turn their directional antennas. Like antennas, rotators come in various shapes

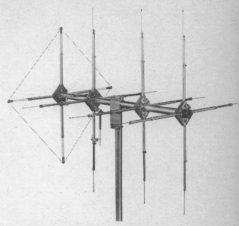

Avanti "Moonraker" can be switched from vertical to horizontal polarization and has a Yagi reflector.

Five-eights wave end-fed vertical base station antenna. The aluminum elements are vinyl covered to reduce rain and snow static. (Anixter-Mark)

An omnidirectional CB base station antenna mounted on a ground mounted tower like this can be as high as 60 feet at its highest point.

When working on any antenna tower a safety belt is a must.

and sizes. The main thing in selecting a rotator is to get one large enough for your needs, both present *and future*. Too small a rotator won't necessarily break down under your load. Instead, it simply won't turn the antenna when there is any appreciable wind or—unless it has an *effective* brake—it will permit the antenna to turn wherever the wind swings it, not where you want it. A *good* TV rotator can handle a small CB beam under most conditions, but if you're putting up a big antenna or your area suffers from chronic high winds, you'd better invest in a good ham radio rotator.

With rare exceptions beams should be tower mounted. This applies to both fixed and rotating arrays. The type of mount adequate to support a simple half wave vertical through the ravages of winter weather just won't hack it with the amount of wind resistance of even a small Yagi or quad. A short rooftop tower is fine—if your roof can take it! If you're not sure, ask somebody who is a carpenter, not your next door neighbor or the CBer down the street.

Whatever antenna you put up and whatever you put it up on, read and follow all instructions. Above all, stay away from power lines with your antenna installations. Dozens if not hundreds of CBers and hams are injured or killed each year when the antenna they're putting up gets away from them and drops onto power lines. This problem has become so acute that power companies in some localities even offer to assist or advise CBers with their antennas. It would be worth a phone call to see if your power company is one of them. When in doubt about your installation, do it better—heavier bolts, additional guy wires and the like. The best antenna in the world is worthless when it's lying on the ground!

BASE STATION ANTENNAS AND THE RULES

Because CB radio is supposed to be a short range service, the FCC has made specific rules limiting the maximum heights of CB base station antennas. For the simplest antenna installations (omnidirectional) the permitted antenna height is 60 feet above ground for the *top* of the antenna if it's on a ground-mounted tower or mast, or 20 feet above the top of the supporting structure (building, *not* a radio tower) if that structure is more than 40 feet high. An omnidirectional antenna mounted on a tower must be mounted so that its highest point does not exceed the 60-foot limit, even if that means mounting it on the side of the tower.

Directional antennas are much more restricted with respect to height. A beam mounted on a tower is limited to 20 feet above ground, and one mounted on a house or other non-tower structure must not exceed 20 feet above the structure on which it is mounted.

FEEDLINES

Your antenna feedline is just as important to the proper functioning of your CB station as the antenna or rig itself. It's the feedline that determines how much of your transmitter's signal gets to the antenna to be radiated, and— looking in the other direction—how much of the other station's signal picked up by your antenna actually gets to your receiver to be heard.

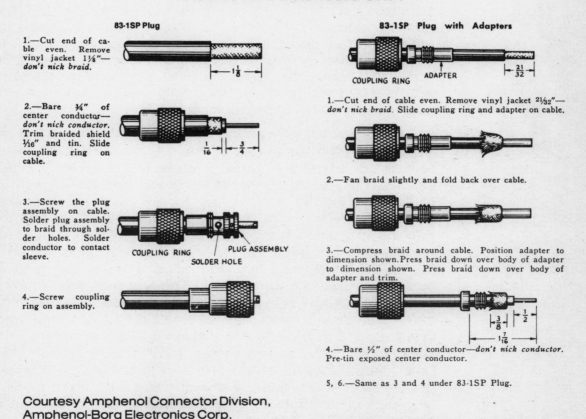

How to wire a standard UHF.

83-1SP Plug

1.—Cut end of cable even. Remove vinyl jacket 1⅛"—*don't nick braid*.

2.—Bare ¾" of center conductor—*don't nick conductor*. Trim braided shield 1/16" and tin. Slide coupling ring on cable.

3.—Screw the plug assembly on cable. Solder plug assembly to braid through solder holes. Solder conductor to contact sleeve.

4.—Screw coupling ring on assembly.

83-1SP Plug with Adapters

1.—Cut end of cable even. Remove vinyl jacket 21/32"—*don't nick braid*. Slide coupling ring and adapter on cable.

2.—Fan braid slightly and fold back over cable.

3.—Compress braid around cable. Position adapter to dimension shown. Press braid down over body of adapter to dimension shown. Press braid down over body of adapter and trim.

4.—Bare ½" of center conductor—*don't nick conductor*. Pre-tin exposed center conductor.

5, 6.—Same as 3 and 4 under 83-1SP Plug.

Courtesy Amphenol Connector Division, Amphenol-Borg Electronics Corp.

All kinds of wire can and have been used for antenna feedlines and to some degree almost anything will work. The question is how well it will work. Any compromise in your feedline will compromise the effectiveness of your station. There is no substitute for good quality *radio frequency* coaxial cable.

All cable has losses, even the heavy nitrogen filled semi-rigid cable used by commercial UHF two-way base stations that costs several dollars per foot. Not only is that kind of cable too rich for our purposes, it's also so thick and stiff that it's a real bear to install. What we need is the best compromise; something we can afford that offers adequate performance and is not too tough to work with. Though there are dozens of different cable types on the market, we're really only looking at two when it comes to CB as well as ham and marine use—RG-8 and RG-58. Both are good. Which is best depends on the application.

RG-58 is the feedline cable used in almost all CB installations. It's about half the cost of RG-8 and—in the typical installation—has so little loss that the difference between the two couldn't be measured without the most sophisticated test equipment. Furthermore, it's less than half the diameter of RG-8, a very significant difference when you're running it under a car seat or through a window. However, if you've got one of those difficult installations that puts the antenna and rig a long distance apart—100 feet or more, you might want to consider the larger cable despite the added expense.

Some words of warning about "RG" cables. RG is the designation formerly used by the military for cables made to their specifications. Unfortunately, there is a lot of coaxial cable now being sold as RG cable that does not perform to the specifications published for RG cables. In general, cable that bears both the manufacturer's name and the cable designation is a good bet. One way to check cable qualilty is to cut back the outer insulation a ½-inch or so. If the shield braid is skimpy and doesn't cover the inner insulation *completely*, don't buy it! The first place a manufacturer cuts corners is in the amount of copper he uses in his cable shields, and less than full shielding means your lead-in will radiate part of your signal instead of conducting it to your antenna. Such line radiation will interfere with your (and your neighbor's) hi-fi and television, and—conversely—inadequately shielded line is much more likely to pick up appliance noise and television "birdies" and disturb your CB reception.

Though coaxial cable looks and is fairly rugged, it can be damaged. Heat is one of the worst enemies of coax, so be sure and keep your feedline away from radiators and hot water pipes in your base station installation. Heat can be even more of a problem in a mobile installation. Stay

You don't have to be high to get hit by lightning. This loading coil was on an antenna mounted on the roof of a mobile home.

well clear of exhaust pipes and catalytic converters and keep your feedline out of the engine compartment if you possibly can. In fact, you will be 'way ahead if you keep the feedline in your mobile installation entirely inside the car—flying rocks and road salt can ruin cable in short order.

Avoid chafing or nicks in the outer insulation. Water can corrode the thin shield wires quite quickly. Avoid sharp bends and pinches like the plague since they can damage shield or center conductor wires or cause a short (immediately or much later) without obvious evidence to the naked eye.

SUMMING UP:

Your antenna is the key element to a good performing CB station. A good quality antenna, properly installed and adjusted according to the manufacturer's instructions, will give you good results for years.

Though this "pincushion" looks impressive, it's little if any better than it would be with only one element.

chapter 6

CB ACCESSORIES

goodies to improve your station performance

The variety of CB accessories that has come on the market has increased almost as fast as the CB service itself. Some of the offerings are very useful—almost necessities. Many of these will be covered in the chapters on security and service. However, there are lots of others, useful and not so useful, that also deserve mention. Those we'll cover here.

FILTERS

Perhaps the most important CB antenna accessory, at least from your neighbor's point of view, is a TVI filter. These come in two types and, though similar in concept, have distinctly different purposes. A *low-pass filter* is used with the transmitter and receives its name from the fact that it passes all frequencies below its cut off frequency with little or no attenuation, while frequencies above the cut off frequency are sharply reduced. All transmitters have some harmonics in their output signals, signals that are multiples of their fundamental output frequency. The second harmonic of a 27 MHz CB signal is just above 54 MHz which is the lower edge of channel 2 TV. The third harmonic of 27 MHz is 81 MHz which is the top end of channel

(Facing page.) In some stubborn TVI cases a line filter like this one may be required. (Avanti)

The showcase at a well-equipped CB or amateur store will have many interesting accessories for CB mobile or base station use.

Use of a low-pass filter keeps harmonic of your transmitter's output signal from getting to your antenna and being radiated into nearby TV and FM receivers. This Telco Products' model can be tuned for minimum interference to a specific channel.

A simple high-pass filter like this Avanti placed in a TV set's feedline prevents your 27 MHz signal from overloading the TV set.

This 12-volt line filter for mobile use keeps alternator whine and other automobile electrical system noises from reaching your CB rig. This filter is about 2½ inches square. (Standard Communications)

5. It doesn't take much signal at one or both of these frequencies to raise hob with TV reception all over your neighborhood.

A good low-pass filter should have no more than a fraction of a dB attenuation at the signal frequency but should have at least 40dB (preferably more) at the harmonic frequencies. Use of a low-pass filter virtually assures you that any TVI problem you experience isn't your fault but the fault of the viewer's receiver. That's where the high-pass filter comes in.

The *high-pass filter* is the converse of the low-pass filter—it attenuates all signals below its cut off frequency while passing higher frequencies with very little loss. High-pass filters are needed because TV set manufacturers cut corners and under-design their sets' tuners. This saves them a few cents in production costs, causing you the aggravation of irate neighbors and your irate neighbor the price of an add-on filter.

Most TVI is caused by properly operating transmitters being picked up by the TV set's feedline and overpowering the set's poorly designed input circuit. The high-pass filter simply shunts the undesired low frequency CB, ham or other signal to ground while it lets TV signals pass.

Line filters prevent still another but less common form of CB interference from spoiling TV reception— interference brought into the TV set by power line pickup. The same kind of filter is also useful for keeping your CB signal from getting into the house wiring through your CB's line cord (unlikely) and for keeping the line cord from bringing electrical noise on the power line into your set. A line filter is often needed in a mobile installation to keep alternator whine, a wavering whistling sound, from getting into a CB set's audio circuits.

Filters are useful for reducing or eliminating many kinds of interference, by the way—electric organs and high-fi systems can often be helped by use of a high-pass, low-pass or line filter. One rule in using a filter: Always connect it to a point in the line as close as possible to the equipment it's being used with. Then follow the manufacturer's instructions exactly to insure that you get maximum benefit from it.

ANTENNA MATCHING DEVICES

In theory antenna matching devices are not necessary, since a properly adjusted antenna in good condition should present a proper match to your transmitter. In the practical world this situation all too often does not exist. Antennas are put up incorrectly, age or become wind damaged. Some are improperly designed in the first place. The transmitter, however, works best only when it is working into the proper load. That's why a matching device is a worthwhile investment, providing it's used correctly (and that means following manufacturer's instructions!), and the antenna problem that made it necessary is corrected as soon as possible.

ANTENNA SWITCHES

If you have more than one antenna—a beam plus an omnidirectional antenna, for example—you should have a simple means for switching them. This means an antenna switch and, since RF is choosy about the kind of circuit path it will follow, a switch specifically designed for use with coaxial feedlines is needed. A number of manufacturers make such switches, and almost any make will suffice. If you have a choice, select the one with the lowest VSWR and lowest loss specifications. If the manufacturer doesn't provide specs, it's still probably OK, but someone who takes the trouble to provide a spec sheet generally takes a bit more pride in his product. By the way, an antenna switch is also an ideal way to ground your antenna during electrical storms.

Another very desirable item for base station use is a static discharge device for the antenna feedline. Though not as effective as an actual grounding switch as described above, it does stay in the line at all times so it cannot be overlooked. Don't expect a switch or any other protective device to prevent damage in an actual lightning strike, however. There's so much energy in lightning that it'll almost certainly damage any radio that's still connected to the antenna system.

POWER MIKES

Power microphones are probably the least understood and most abused of all CB accessories. In the first place, a well designed transmitter should not require one since its modulator will be designed to deliver 100 percent or nearly 100 percent modulated signal with the standard equipment mike.

A power mike does increase the average modulation level, to be sure, but to do so it also brings up background noise and changes the operator's voice characteristics. Properly used—and that means adjusted carefully for the user, the radio and the operating environment—a power

An antenna matching device can be very handy if your antenna's tuning isn't quite as good as it should be. (Avanti)

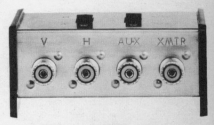

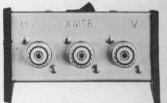

Antenna switches are a must if your base station uses more than one antenna. (Avanti)

Some sort of lightning protection is a must for every base station. Arresters like this one bleed off static build-up, protect your equipment and you too.

Accessory base station mikes offer more features than those usually supplied as original equipment and are almost a must when a mobile rig is used as a base station. This example from Midland has a locking push-to-talk switch and built-in pre-amplifier.

Here's a mobile radio being used as a base with a desk type mike and auxiliary AC power supply. (Standard Communications)

Sharp's base station adapter is designed to have the mobile transceiver it powers mounted on its top surface.

mike can improve signal intelligibility under certain conditions. Improperly used, which if on-the-air observations are accurate is 90 percent of the time, power mikes so distort the signal that it's hard (sometimes impossible!) to read, and the distortion spreads the signal out over several adjacent channels.

The problem of power mike abuse has become so severe that the FCC's chief engineer has been considering some kind of regulation of their use. The specifications under which the new 40-channel transceivers must be tested include some power mike tests.

If you have a *good* reason to use a power mike do . . . but be sure you set it up properly, preferably with the help of a qualified technician and instruments. If you don't really need one, put your money in a more useful accessory.

CONVERTERS

Though not strictly an accessory like those items we've discussed so far, a CB converter can be a most useful automobile accessory for the traveler or even for someone casually interested in Citizens Band radio.

CB converters are compact devices designed to be connected between your car broadcast antenna and the broadcast receiver. When it's turned off, it does nothing, and the car radio performs normally. When it's turned on, however, it converts the 27 MHz CB signals, picked up by the car antenna, down to broadcast band frequencies where they can be tuned in on the BC set.

There are two types of converters being offered, tunable and fixed-tuned. The fixed-tuned type is easier to use in cramped quarters since the actual tuning is done with the BC set. Its big disadvantage is that strong broadcast signals may bleed through and block out CB signals on a channel you'd like to hear.

Tunable converters have their own tuning knob and dial, with the user first setting the broadcast receiver to a predetermined spot on the dial, and the CB channels then tuned on the converter. In general this type works better and is easier to set on a specific channel, say channel 19, but it will cost more. Neither type works as well as a CB transceiver—for example, there's no squelch on most makes so there's always background noise when you're not hearing a signal—but they do offer an access to "instant" road information without the expense or necessity of becoming a full fledged CBer.

OTHER CB ACCESSORIES

There seems to be no end to the CB accessory market, and many useful items to improve your CB activities can be found there. External speakers are almost a must for those transceivers with tiny and/or badly located speakers. Earphones, with or without a built-in mike, are very handy for both base station and mobile use when others who might be irritated by the chatter are present.

Well equipped CB and amateur radio shops usually have a good selection of such items, which are also advertised in the various CB periodicals. As with any purchase, weigh your need and the apparent quality of the maker's effort before you lay your money on the counter.

This base station power supply includes a built-in speaker as well as mounting brackets for the mobile transceiver it powers. (Metro Sound)

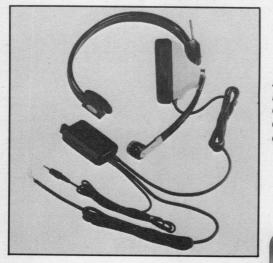
A combination headset/microphone is both convenient and avoids distrubing others nearby when operating from mobile or base.

Few manufacturers make CB receiver converters as fancy as this Tenna which has both squelch and automatic noise limiter.

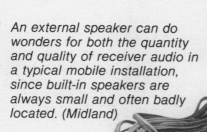
An external speaker can do wonders for both the quantity and quality of receiver audio in a typical mobile installation, since built-in speakers are always small and often badly located. (Midland)

Perhaps the ultimate for CB listening is this Boman BM-1129, an in-dash receiver that also includes AM and FM broadcast and 8-track tape player.

Receiver preamplifiers can often improve weak signal reception. This Telco has an RF-sensing detector which automatically switches it out of the line when the transceiver transmitter is keyed.

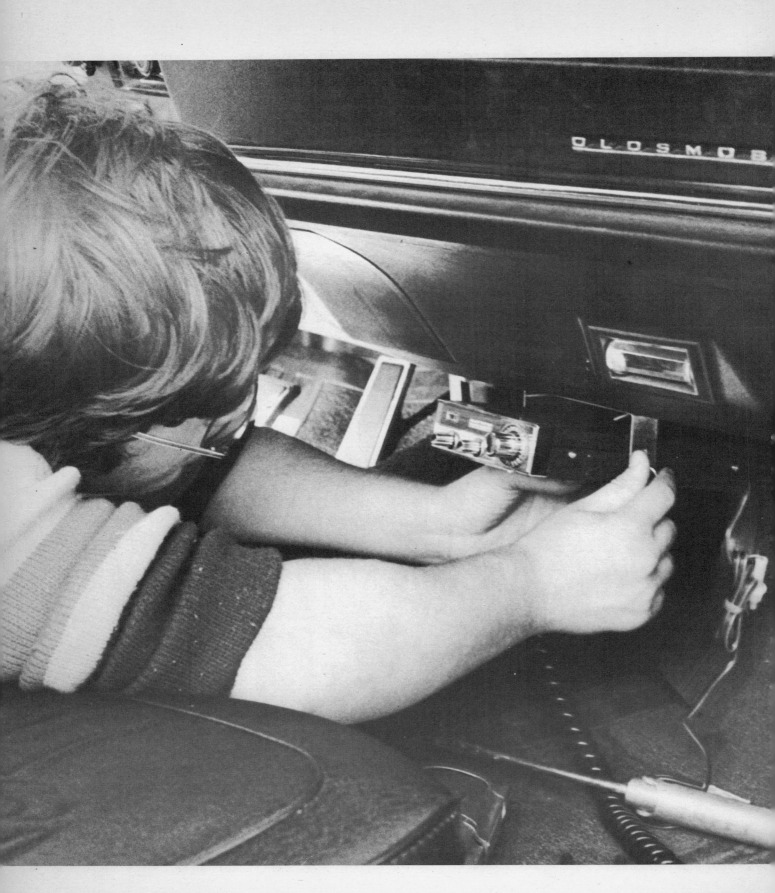

chapter 7

INSTALLING YOUR NEW CB RIG

if you are going to use your radio, it's got to be where you can use it.

To enjoy the many benefits the Citizens Radio Service can provide, you've got to be able to *use* your CB radio. This is not always as simple to accomplish as it sounds, because a poorly laid out installation can not only be inconvenient to the operator but also irritating or even dangerous to family, friends or the operator himself.

Of course, proper installation is more of a consideration in the case of a mobile station than a base because of the reduced amount of space in which to work, the consideration of driving safety (primary task) vs operation of the CB (secondary task) and the ever escalating rip-off problem. But even the CB base station operator should think out what he's doing before he designates a specific room as his "shack" and starts punching holes in walls and window sashes.

Base or mobile, there are three fundamental considerations you must keep in mind for your installation:
- You must be able to see the rig
- You must be able to reach the rig
- Its installation must interfere as little as possible with other activities

By far the majority of CB mobile rigs are simply bolted under the dash like this Standard. Note that the radio is far enough to the right that the mike coil cord clears the steering wheel.

In trucks and campers it's often more convenient to mount the radio on the cab roof. (REACT photo)

Satisfy those requirements and you'll get the most satisfaction from your CB operation.

MOBILE INSTALLATIONS

Usability is the key to a successful mobile radio installation. With a channel selector and other controls you'll want to monitor, your radio pretty much has to be somewhere where you can see its front panel from the driver's seat. Unless you do all or most of your driving solo, it's also desirable to have it located so it's usable from the passenger seat. It's no fun driving through heavy traffic in a downpour with a microphone in your hand asking for road directions, so it's nice to be able to turn the radio operating/navigating task over to the occupant in the right-hand seat under such conditions.

Of course, it's not only necessary that the operator be able to *see* the radio, but he's also got to be able to *reach* it for channel selection or other adjustments in order to operate it satisfactorily. The requirements of both accessibility and visibility are satisfied by hump, under-dash and in-dash locations. However, some CBers have mounted smaller rigs on top of the dash or even on the ceiling—a location preferred by many truck drivers whose cabs are well-suited for such mounting. From the safety standpoint, the closer the radio is to eye level the safer a glance at it becomes. There's hardly a CB mobile operator going who hasn't swung his eyes back up from the rig to find a rapidly growing pair of tail-lights glaring at him! This is one of the problems the new combination microphone/control head operated rigs avoid nicely.

Along with the visibility and accessibility requirements there are some rather basic mobile installation problems that are all too often overlooked—particularly by the newcomer to CB who's installing his first mobile rig. Number one is to avoid letting the mike cable foul the steering wheel. Though a mounting location immediately alongside the steering column might seem pretty ideal from most viewpoints, it's a real disaster the first time you try holding the mike to your lips while you're turning a corner! The coil cord will neatly grip the rim of the wheel as you begin your turn, and if you've done it just right, continuing the turn may even extract the cable from the connector where it's screwed into the rig's front panel with practically no effort on your part! The solution is a simple one. Before settling on your final mounting point, have someone hold the radio in the place you think you'd like it to be while you sit in a normal driving position with the mike to your lips. If the mike cable doesn't clear the wheel, don't mount the rig there!

The second problem is also a mechanical interference problem—that of avoiding blocking the flow of heater or air conditioner air. There are two very good reasons why you should avoid this problem. First, electronic equipment does not thrive on temperature extremes, particularly high temperatures. A good car or truck heater puts out a lot of pretty hot air on a bitter cold day, and letting your CB radio soak up a good portion of that heat will deteriorate its performance at best and may quite possibly lead to premature failure of otherwise long-lived components. The second reason is your own personal comfort. Modern automotive air circulation systems are pretty complex and depend on a lot of outlets to keep all riders comfortable. Don't let a thoughtless two-way radio installation make your accelerator pedal foot into an ice cube whenever the temperature goes below freezing.

MOUNTING BRACKETS

The mounting brackets supplied with CB radios vary from very good to just barely adequate. A good bracket should provide some flexibility for the installer, both in the number and the location of holes for mounting the bracket and in the angle in which the radio mounts in the bracket.

The bracket should also be made of heavy enough steel that it not only continues to hold the radio in place over rough railroad crossings but will also resist the yanking of a strong-arm rip-off artist. Mounting brackets should also be equally usable on the bottom as well as the top of the radio, for dash top or hump installations.

When installing a CB set under the dashboard of an automobile, be sure to select a location that will actually support the weight of the radio. Some current cars have so much plastic in the dash assembly it's actually difficult if not impossible to find a suitable spot to mount a radio. If you're very lucky, you may actually find a suitable empty hole or two in a steel member at the bottom of the dash in just the place you wanted the radio to go. If you don't, you'll have to get out the drill. Just be sure before you drill that you're drilling the right size hole, it's where you want it to be, and when the drill goes through, it's not going to punch a hole in the clock or your car's broadcast receiver!

Mounting bolts should be at least 8-32, but ¼-inch is better. Small screws break, tear through metal and work loose much easier than large ones. Use lock washers—it

Sturdy mounting brackets like these are a real plus when you're selecting a radio. They should also be reversible—mount on either top or bottom of the cabinet—and hold the radio with hardware sturdy enough that bumps and vibrations won't work them loose. Note the difference in mounting slot configurations between the Cobra, Bohsei and Clarion sets shown.

A slide mount as used on this Cobra offers the advantage of quick removal for security. It's also a necessity if the same radio is to be used as a base station or changed frequently from one car to another.

RCA's CB slide mount has a quick release latch and is designed to be used with the CB set's original equipment mounting bracket. Note the heavy duty electrical contacts, very important to avoid signal loss.

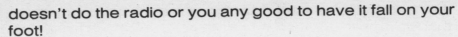

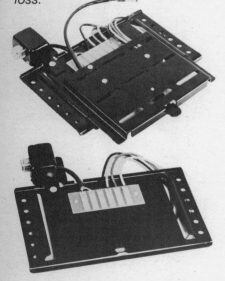

An in-dash CB installation of a set like this Boman is a major project. However, once accomplished, it is not only very neat but practically rip-off proof.

doesn't do the radio or you any good to have it fall on your foot!

As an alternative to the mounting brackets supplied with the radio, you might want to consider one of the various types of accessory under-dash mounting brackets offered by a number of manufacturers. The slide-in mount, originally developed for stereo tape decks when they were the rip-off artists' favorite target a few years ago, is one type, while the heavy gauge lock type mount designed to protect the radio from a thief's attack is the other. Both have their points, but since they're more appropriately discussed in detail in the chapter on security (Chapter 8), let's go on to other mounting styles.

In-dash mounting for your CB rig is a lot of trouble, but when properly done, it can result in a neat, convenient and generally less rip-off prone installation than any other approach. Some commercial CB radios designed for in-dash mounting are available and are described in the equipment section, and as more and more car manufacturers offer CB sets as factory equipment, in-dash mounting will become more common. At the moment, at least, in-dash commercial CB radios are a good deal more expensive than conventional sets. How practical is it to make your own in-dash radio installation?

For the mechanically minded CBer with access to metal working tools, a home brew in-dash mounting job is certainly a possibility. The first and foremost question is whether there's a place in your car for it. Panel space (and remember the "usability" we emphasized with respect to location) is only the beginning. Most modern cars have so much ductwork and so many control mechanisms behind their dashboards that it's tough to find a spot to accommodate even a super compact rig—and don't forget rear clearance for the antenna connector!

Assuming you've got a spot that works, cut a hole to clear the radio and frame it with either a commercial BC radio bezel (available from some auto supply stores or the larger radio hobbyist outlets) or make up your own out of aluminum or brass molding stock. As for supporting the set behind the panel, a shelf or a hanging clamp made of perforated strip stock will work, or . . . you'll have to work up something appropriate for your situation. Some cars and vans have removable dashboard panels that really lend themselves to a do-it-yourself in-dash installation. If you're fortunate enough to have one of these, the problem becomes much easier.

Another approach to an "in-dash" mounting is to put

the radio into your car's glove compartment. It's an excellent approach from the security standpoint, since it's completely hidden when you're not using it and the door is closed. However it's not very convenient since the glove compartment is usually well to the right and your passenger may not appreciate the glove compartment door hanging open much of the time. Nonetheless, some very nice semi-permanent CB installations have been built behind glove compartment doors.

A very good location for a built-in CB set is the between-the-seats console found in many bucket seat equipped cars. Such an installation often lends itself to at least partial concealment, too. Since variety is the common denominator when it comes to console design, you're on your own as for details. The console is pretty ideal for operating convenience. Just remember, though, that any type of console as well as in-dash mounting is going to mask the set's built-in speaker. You're going to have to provide an auxiliary speaker and a place to mount it so you can hear the radio.

Hump mounting is a good answer—sometimes the *only* answer—for cars without consoles. Some hump mounts are simply mounts, often designed to be used with the radio's standard mounting bracket. They may or may not include an auxiliary speaker. One type even channels the sound from the CB set's own bottom-mounted speaker through an internal baffle which directs sound toward the user. Hump mounting, like console mounting, is very convenient and an alternative well worth considering.

In its simplest form the hump mount is nothing more than a piece of sheet steel straddling the hump. The CBer with access to sheet metal working facilities might even want to try working up his own. It's even possible with ingenuity to make one from wood.

As often as not, commercial hump mounts for CB radios offer some sort of security latching. This type is covered in depth in the security chapter.

ELECTRICAL HOOKUP—POWER

First consideration in making your mobile radio electrical installation in your automobile is the power connections, and the primary rule to remember here is *always observe polarity*. Red connects to red, and black goes to black (or ground—the car chassis—in most cases). Though there are a few exotic cars around which have the positive (plus) battery terminal grounded, most of the cars on the road

A home brew in-dash installation. The owner of this camper made use of an existing panel cutout to come up with this neat job.

This couple solved the problem of mounting a Standard CB set in their van by mounting it face-up in a writing board.

Hump mounting this Hy-Gain with its original equipment mounting bracket makes a functional installation in this car.

80 Installation

When room permits, a special hump mount like this Gambler-Johnson with built-in speaker makes a neater, more functional installation.

In tight situations like this sports car the hump provided the only practical mounting for the Hy-Gain transceiver.

Transmission humps make excellent locations for CB rigs. This Johnson 123SJ is installed in a Magni-Power locking mount.

have a negative ground electrical system. If you're driving a positive ground vehicle, be sure the radio you're planning to install will work with positive ground (a good many will but not all) and follow the hookup instructions you'll find in the radio's owner's manual. If you're not sure whether you have positive or negative ground, *find out* before proceeding further. Your car owner's manual or the service department of your auto dealership should be able to supply the answer.

You have several options in the choice of tie points to tap into your car's 12-volt electrical system. The worst choice is (usually) a cigarette lighter plug, while the best is a pair of husky wires running through the firewall and screwed directly to the clamps on the battery cables. Most installers end up with a compromise, wiring the hot lead into the automobile's fuse block and grounding the negative lead to some nearby screw on the frame. Whichever point you choose, be sure to use sufficiently large wire. Number 16 is adequate, but No. 14 is better.

The cigarette lighter plug is a poor choice primarily because it's mechanically unsound! Some lighter plugs are so badly made that their internal electrical resistance will drop the voltage to your set excessively. The net result to you is reduced receiver sensitivity and less speaker power on receive, lower power output on transmit, and in extreme cases the rig may not function at all! Even the best cigarette lighter plugs can work themselves loose causing the radio to go off the air, and dangling wires drooping across the dash certainly don't add to the car's decor. The cigarette lighter plug's only plus is a logistical one. If you do a lot of operating from rented or company cars where it's not practical to install permanent wiring, your only choices may be the cigarette lighter plug or spring-loaded clip-on connectors (alligator clips) on your power leads. Since alligator clips can pull loose easily, can also move around on their terminals and can short circuit to nearby hardware, cigarette lighter plugs are usually a better choice for a temporary installation until you find yourself in an economy car with no cigarette lighter!

The big advantage of going directly to the battery is voltage stability. Though the voltage in your car's electrical system varies widely—from less than 10 volts when the starter's cranking a cold engine to 15 volts or so at full charge—the variation is least at the battery since it provides the regulating reference of the electrical generating system. The battery also provides another benefit, that of filtering out noise that might be generated by the car's electrical system so it doesn't get into the CB rig. Alternators are particularly bad offenders in this respect, and "alternator whine"—a warbling or whistling sound that varies in pitch as the engine speed changes—can get into both the receiver (where it's heard in the background of every received signal) and transmitter (where it modulates the carrier along with your voice) of a CB rig connected to the car's wiring at a point well removed from the battery. However, most CBers don't believe the benefits of running extra wiring up into the engine room is worth the trouble it takes—and in most cases they're right! The fuse block is usually a good enough choice, and it does offer other advantages that shouldn't be overlooked.

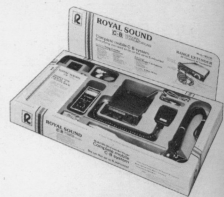

Manufacturers and sometimes dealers offer complete packages like this one from Royal Sound. It includes antenna bridge, cigarette lighter plug and even license forms as well as transceiver and antenna plus promises to put the purchaser "on the air in 5 minutes"—it probably comes pretty close to that.

Wiring your CB set into the fuse block offers a flexibility that no other tie point does. In the first place the fuse block is designed to be the central tie point for all your electrical accessories so it can supply ample current, has built-in fuses and is relatively convenient. Perhaps its biggest advantage is flexibility. By selecting the proper point on the fuse block, you can set up your wiring so the radio will operate only when the ignition is on, when the ignition switch is switched to ignition or accessories, or at any time. The first two options are a good idea if you have small children in your family or you're the forgetful type that might walk away and leave the set running. Unless you or someone else with a key is in the car, your radio is dead, not further cluttering the crowded band as a child's plaything or slowly running your battery down.

Selecting the desired tie point on your car fuse block requires a little bit of exploration but isn't too tough a job if you have a helper. Turn on the set, set the volume control about halfway up and turn the squelch control fully counterclockwise. Ground the negative power lead (black) to a convenient point on the chassis and then start touching (careful!) the positive lead to various exposed terminals on the fuse block. With the ignition off you'll quickly determine which points are "hot" (loud rushing noise from the speaker) and which are not. Then have your helper turn the ignition switch to "accessories" and note which of the formerly dead terminals are now "hot." Finally, have him turn the ignition switch to "on" (but don't bother starting the engine) and touch the remaining terminals. You now know which are which.

Depending on the make of car, the fuse block may or may not have some kind of terminals available to fasten your power lead to. Since fuse blocks are designed to be the tie point for nearly every possible factory electrical option available, you'll almost invariably find some open slots on it that you can use for your radio. You'll have to add a fuse to the slot you select—5 amps should be sufficient—but then you should always fuse your wiring anyway. In some cases unused slots on the fuse block won't have a fuse clip at the accessory end, but this can actually make using that slot easier. Simply buy a pair of snap-on fuse clips at your hardware or radio store, solder the hot lead from your radio to one of the clips and snap the clip on one end of the fuse. Then snap the other end of the fuse into the fuse clip in the fuse block and presto!—you're all set to go! Still another "quick and dirty" is to use one of the recently developed fuse block "cheaters"—little lug-like terminals that slip under one side of an already-installed fuse on the fuse block to provide a power take-off point. Most well-equipped CB dealers should have them in their stock of accessories.

Some final words of caution about wiring. Always route wiring out of the way where it can't get caught in your feet when you're getting in or out of the car and where it won't get tangled in the floorboard pedals. Loose wiring can be dangerous, so it's always a good idea to support wiring by taping it to appropriate behind the dash supports at frequent intervals. Fuse your set's hot lead. The manufacturer of the radio usually does, but if he doesn't, you must for the protection of the radio, the car and yourself. Finally, install a plug and socket somewhere in the power leads if the manufacturer hasn't provided one. Even if your car is always used in areas where rip-offs are unheard of (if there is such an area!), there'll be times that you'll want to take your CB radio out of the car for service, for use as a temporary base station or whatever. Jones or similar polarized plugs (so you can't push the plug in with ground and hot leads reversed) are inexpensive and make a convenient means for a common hookup system in different cars or on the bench. Just remember that the female connector is used for the car (hot) end of the circuit while the male plug goes to the radio. Exposed hot terminals in a temporarily

"Always route wiring out of the way where it can't get caught in your feet when you're getting in or out of the car and where it won't get tangled in the floorboard pedals. Loose wiring can be dangerous."

The first step in making a CB mobile installation is to find a mechanically suitable mounting location that is convenient to the driver. This is most easily done with a helper.

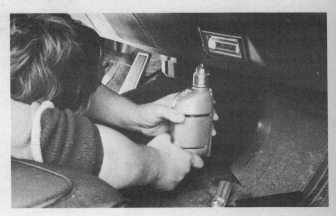

Use the mounting bracket to locate the mounting bolt holes, then drill the holes taking care not to damage car wiring or accessories behind the dash with the drill bit.

After fastening the mounting bracket to the dash with suitable bolts, mount the radio in the bracket and adjust it to a comfortable angle for viewing and operating.

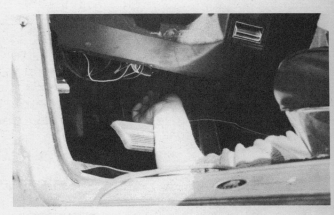

Thread power and antenna leads over the steering column in such a way that they don't droop down and get tangled with your feet or the brake or accelerator pedals.

Locate a screw or small bolt that is screwed into a *metal* portion of the car frame and fasten the negative power lead to it.

Connect the positive power lead to an appropriate spot in the fuse block. For safety's sake, it's best to disconnect the positive lead to the car battery when you're working around the fuse block.

Trunk lip mount antennas are easily installed, neat appearing and work very well. The first step in installing one is to position the mount bracket snugly against the edge of the trunk lip.

*Draw the set screws **tight** with the Allen wrench provided. The tips of the screws must break through the paint in order to provide an effective ground for the antenna.*

Install the loading coil and spring (if any) on the base assembly. Tighten it well so vibration won't loosen it.

unused plug can provide plenty of excitement when bouncing around behind your dash!

SPEAKER WIRING

The only other wiring you're likely to need for a CB mobile installation is for speakers—an accessory speaker to replace the CB set's built-in speaker and an outside speaker for PA use if your set has that option and you want to use it. With CB sets being made as compact as possible, the built-in speakers are often just barely adequate, either in size or location. An accessory speaker, either separate or built into a hump mount for the radio, can bring a remarkable improvement in both quantity and clarity of received signals.

In general the same precautions apply to speaker wiring that were already covered under power and antenna cabling. Speaker wires should be routed out of the way, and the speaker itself mounted in a location where it doesn't impede any air ducts yet still directs its output in the direction of the listener's ear. For PA use get a speaker made for that rough service. It has to be weather- and waterproof. Use fairly heavy wire for speaker hookup— No. 18 is adequate, but 16 is better.

ELECTRICAL HOOKUP—ANTENNA

The mechanical aspects of mobile antenna installation were covered in the chapter on antennas, so here we'll concern ourselves only with electrical considerations. We've got our antenna mounted—trunk lip, rooftop, rear fender or whatever—and now our prime concern is running that length of coaxial cable from where it's attached to the antenna to the back of the rig.

Wires do not like heat, pressure or abrasion. If at all possible, *never* run wiring under your car. Heat from the exhaust system (catalytic converters get so hot they can ignite grass!) will destroy coax in no time, while nicks in the outer insulation will let water get into the shield braid where it will corrode and drastically decrease cable life. Pinched coax, even though it doesn't fail immediately, will eventually short, putting you off the air and possibly even causing transmitter damage. Watch out for trunk lip mounts in this respect. You may have to put a slight crimp in the tongue and groove along the edge that seals out moisture or you'll end up with flat, useless cable right at the point where it comes out of the mount.

Routing the antenna cable from trunk to dashboard is

not too tough a task if you think it out first. Be sure to run the cable along the sides or up high in the trunk so it doesn't get snagged and broken when you're loading the trunk. Be careful running it under the rear seat, since the rear seat often rests on a heavy wire frame that would act as an efficient wire cutter when supporting a couple of your husky football player friends. Getting the cable from the back seat to beneath the dash can usually be done by loosening the molding at the bottom of the door that holds the edge of the carpet and slipping the cable under the carpet edge or under the plate itself. From there on you're home free—as long as you don't let cable slack dangle down so it tangles with the driver's feet! Running the cable alongside the transmission hump is still another good option.

Insert the whip in the antenna collar and tighten the set screw. NOTE: Most antennas will require tuning with an antenna bridge to operate with maximum efficiency.

Roof mounted antenna installations are a lot more sticky and in many cases call for professional help. Mounting the antenna directly above or just fore or aft of the dome light gives you access to the antenna base, but you'll still have to fish the feedline under the headliner and down one of the window posts. Experiment before you cut that hole in the roof. If you can't find a way to run your feedline where it has to go by yourself, you're either going to have to seek professional help or go to a different type of antenna.

If you're determined not to cut any holes in your car, a gutter clip antenna mount will still get the antenna high in the air. (Antenna Specialists)

Whatever your choice of antenna location, once you've got it and your power cable is connected to the back of the rig, you still can't relax. Your electrical installation job isn't finished yet!

NOISE CHASING

Turn on your rig and flip the channel switch. You should hear someone talking on at least some channels unless you're an awfully long way out in the boondocks. Select a channel that has a weak but readable signal as indicated by a low reading on the S-meter* or audible noise riding in on the background of the signal. If the set has a noise limiter or noise blanker switch, turn it off. Now start your engine. Do you still hear that weak signal? If so, you're very lucky. Most automobiles are terrific noise generators. If yours is typical, you can't tell if that weak signal exists while the engine's running!

Now switch on your noise blanker or limiter. If it's at all effective, the noise level should drop off, and you might even be able to hear that weak signal still talking

---

*The S-meter is another name for the front panel signal strength meter found on most CB sets.

Installation

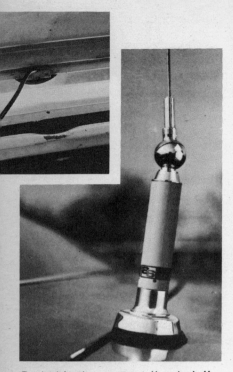

Probably the neatest "no hole" mobile antenna mounting is a trunklip mount like this from Antenna Specialists. It's sometimes necessary to cut or crimp the rain groove slightly (inset) to avoid pinching the coax feedline.

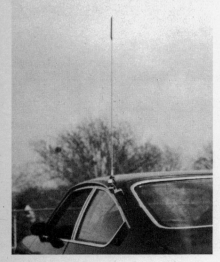

For hatchback drivers a gutter clamp mount like this one from Antenna Specialists can do the trick.

underneath the noise. Race your engine a little bit, however, and it'll probably disappear again in the increased noise. There's an old ham radio adage that says, "you can't work 'em if you can't hear 'em!" Clearly something needs to be done if you're going to start hearing 'em while on the move.

Ignition noise—the electrical impulses that fire the spark plugs that make your engine run—is the prime contributor to car electrical noise. It's radiated by your engine's ignition harness and picked up and re-radiated by other wiring and ungrounded metal parts throughout the car. Usually, but not always, you'll find that ignition noise is getting into your receiver from the antenna. There's an easy way to find out—simply unscrew the antenna from its mount and see if the noise goes away. If it doesn't, you've got a stickier problem, but let's take the most common case first. You removed the antenna and the noise dropped drastically or disappeared entirely.

Before proceeding further with our noise hunt, it's time to assemble a simple but most effective noise fighting tool, a grounding lead. Take a foot or so of fairly heavy flexible stranded wire—No. 14 should be fine—and fasten a husky jawed clip such as an "alligator" clip to either end. Reinstall your antenna, turn off your noise limiter and blanker, turn up the receiver volume so you can hear it outside the car, and start the engine. You should hear the staccato pop of ignition noise, so you're ready to begin the hunt.

The first place to look for a noise-radiating antenna is under the rear of the car—your tail pipe which, after all, is a metal tube somewhere between a quarter and a half wave in length. The tail pipe is usually hung by rubber insulating blocks and is directly connected to that big, powerful noise generator, the engine. Be prepared to get a little dirty, by the way—auto noise chasing is not a clean job! Crawl under the rear bumper with a file and your grounding lead, scrape a bare spot on the end of the tail pipe with your file (rust, road dirt and tail pipe soot are all good insulators), and connect the tail pipe end to a good chassis ground (you may need to insure a good connection here with a few strokes of the file, too) and listen to the receiver noise. Did it decrease? If so, you're going to have to drill a small hole near the end of the tail pipe and another nearby on a frame member and then connect the two with a piece of copper braid (leave a little slack) fastened down with self-tapping screws.

Similar experiments should be performed on the hood, engine, trunk lid and both bumpers. Any ground that

RECEIVER NOISE SOURCES IN MOBILE INSTALLATIONS

Receiver Hears	Probable Source	Cures
Steady popping that varies with engine speed.	Ignition system.	1. Bond (in order of importance) tail pipe, hood, trunk lid and bumpers to car frame with strips of copper braid. 2. Have an ignition system tuneup. 3. Replace ignition harness with new factory specified resistance wire.
Irregular popping, most noticeable at higher engine speeds.	Voltage regulator.	1. Place a high current coaxial capacitor in series with the battery lead at the regulator.
Whistling or whining, pitch changing with engine speed.	Alternator.	1. Add an alternator filter in series with the power lead to the CB.
Irregular clicking or crackling noises while standing still, engine running.	Electrical Gauges.	1. Add bypass capacitors or tuned filters to guage wires.
Electrical motor noises.	Heater or windshield wiper motors.	1. Place a high current coaxial capacitor in series with the motor power lead, right at the motor.
Irregular popping only when car is moving.	Wheel or tire static buildup.	1. Add static-collector springs to front wheel grease caps. 2. Add anti-static powder to each tire (don't forget the spare!)

reduces noise should be made permanent with a short piece of flexible wire or braid. It's a bit of a dirty job and one that's somewhat tedious, but it's one that can have a dramatic effect on mobile reception and is well worth the trouble. Don't fail to do it!

If grounding doesn't do much for your reception, and you've got almost as much ignition noise when your antenna is off as when it's on, you're suffering from *conducted* interference. It's coming through the car wiring and into your receiver through the power cable. Commercial filters that go in the power leads may clean up the problem, and wiring the rig directly to the battery cable terminals instead of the fuse block should also help. If conducted interference is your problem, good luck. It may take a lot of fooling around with shields and by-pass capacitors if you want a really quiet mobile installation!

BASE STATION INSTALLATIONS

Actually, most of the same considerations apply to a CB base station installation that we've already discussed earlier in this chapter. The most significant difference is that you've (usually) got a lot more room to work in. The other difference is that you've usually got other people—family or business associates—who need to be considered before you put the finishing touches on your CB base.

Before you decide on what part of what room or on what floor you install your station, you have to ask yourself the following questions: Who's going to be using the radio? What time(s) of day will it be used the most? Who is going to be bothered by CB chatter? Answer those questions and you're well on the way to sensible station arrangement.

An effective CB base station can be set up wherever you want it to be. This bedroom installation consists of a Pace 1000M AM/SSB mobile transceiver powered by an NPC 104R AC supply.

Perhaps the best way to approach the answers to those questions is with a couple of don'ts.

- Don't put the station where it's inconvenient or inaccessible to those who will be using it regularly.
- Don't put the station where its use or simply existence bothers others.

In other words, plan what you're doing before you do it.

Those aren't the only considerations, of course, though they are primary. You've got to get the antenna lead-in to the radio which means you're better off near an outside window. When it comes to getting the lead-in to the radio, that's not too tough a challenge. The neatest way is to cut a small hole just large enough to clear the size

This CBer's Hy-Gain III takes very little room in his well-filled roll-top desk.

coaxial cable you're using in the window jamb. When and if you want to plug it up, you can do so easily with a piece of dowel and a dab of paint. If you can't quite bring yourself to drilling that hole, there's another way. Just run the coax in through the top of the window, lay a piece of foam rubber weather strip across the top of the window and close it.

Second, if you've got a choice you're much better off locating the rig near a telephone. Even if your first interest in CBing is not emergency or assistance work, sooner or later you're going to find yourself using the two at the same time. If they are at opposite ends of the house or office, that isn't going to be very convenient.

If you're using a mobile rig with an auxiliary AC power supply, it's a good idea not to have the two too close together. Power supply transformers generate a strong electric field that can be picked up by nearby sensitive amplifiers, and many a CBer has found he could solve a mysterious hum problem in his transmitter and/or receiver simply by taking his transceiver off the top of the power supply and putting it several inches away. Of course, this is rarely a problem with the accessory AC supplies made by CB manufacturers for use with their own transceivers, but if you do have a hum problem with your mobile rig when using it as a base, this is a prime suspect.

There are some very useful accessories worth your consideration in your base station installation. An antenna low-pass filter, feed line grounding device and antenna switch (if you have more than one antenna at your base) are practically musts. Take a look at the previous chapter on accessories for some other ideas.

Use good sense and consider the needs of both yourself and others when you plan your base station installation—the extra time you invest before you set it up will pay dividends as you use it.

Base station CB installations offer many more options and much fewer problems than their mobile counterparts. Putting the station in the kitchen placed it near the center of activity, convenient to both telephone and windows for antenna lead-ins.

chapter 8

SECURITY

if you're going to use your radio, you'd better not lose it!

In many areas just hanging on to a CB radio or any other electronic accessory that's mounted in a car has become a major problem. In one eastern city a high police official estimates that a CB set in a car driven and parked on the streets of his city will last less than 3 weeks. In some areas of almost any big city the figure might well be closer to 3 hours! The rip-off problem has become so acute that a number of states are permitting insurance companies to exclude CB radios from their comprehensive coverage, so even if you're presently covered, the chances are very good that you won't be the next time you renew. What can you do, then, to try and keep your rig for yourself?

First and most obvious is to never leave your car unlocked. CB radio thieves are like other thieves: they prefer to work quickly, and the easier you make it for them to take your radio, the more likely you are to become their victim. Thieves prefer to work in private, too. A car parked halfway down a dark alley is far more likely to be hit than one under a street light next to a busy sidewalk.

The tranquillity of suburbia is no guarantee of security, either, as radios disappear from cars parked in streets, driveways and even garages. Don't be lulled by a false

(Facing page.) Lock mounts of this type are less convenient than slide-in mounts since electrical connections must be made individually when the rig is installed or removed. As a general rule, they are a little tougher for the rip-off artist to conquer.

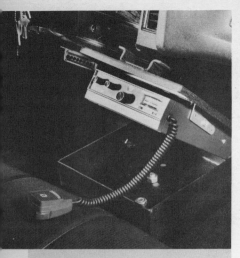

If you're going to leave your rig in the car, there's no way you can absolutely guarantee its security, but the "CB Safe" comes close. Very heavy bolts fasten it securely to the car's floor while a special padlock protects its contents when the rig's not in use. Cable access is a rear panel hole. (F.H.S., Inc.)

sense of security in a public spot. Next time you're at a big shopping center or even a supermarket parking lot, notice how little attention anyone pays to a person repairing "his" automobile. He could just be a gutsy thief busily stripping it of a CB and other expensive accessories while the real owner is off shopping. Is there anything we can do to put off this kind of thief?

First it should be pointed out that if someone is absolutely determined to steal a radio from your car there is no possible way you can stop him from doing so. What you *can* do is slow him down and attract attention to what he's doing. If you do these things well, he may decide to go for easier prey.

MOUNTING FOR SECURITY

More secure mounts are one answer. Most CB set mounting brackets are simple sheet metal affairs designed to be bolted to the bottom edge of the dashboard where they hold the radio in a convenient position for the driver's operation. The radio mounts to the bracket with thumbscrews, a simple, easy-to-install arrangement that makes it equally easy for even the most inept thief to have your radio out of its mounts and under his coat in a matter of seconds.

The simplest way to make a standard mounting more resistant to rip-off artists is to replace the mounting thumbscrews with one of the several types of special "secure" screws available at many CB shops and hardware stores. These require a special tool to unscrew so they can provide a fairly effective and inexpensive deterrent to less sophisticated auto burglars, but they do have several obvious disadvantages. First and most important is that they aren't all that effective against a better equipped or more ambitious thief who is probably carrying the special tools needed to unscrew them, or lacking the tools or the patience to use them, will pop the mounting bracket off the dash with a pry bar. Special screws can also be very frustrating to the rig's owner when he wants to take his rig out of the car and the tool is in the house or misplaced.

A special lock that fastens over the head of one (or more) of the mounting thumbscrews is no more effective than the special screws. Mounting thumbscrews are typically light enough that a sharp rap on the side of the lock with a screwdriver handle or pair of pliers will break them right off, so the thief has only to unscrew the remaining unprotected screws and he's off and running with your radio.

A better solution is one of the special locking type mounting brackets that are now being widely offered. These are generally made of much heavier steel than the standard mounting brackets and are adjustable to hold rigs of various sizes securely while resisting attack by screwdrivers, pliers, a pry bar or whatever. Mounted under the edge of the dashboard, a heavy-duty locking mount may truly protect your rig—but don't be too sure! Some current auto dashes are so flimsy that rig thieves have been making off with rig, mount—and a good chunk of the car they were fastened to! The thief retires to a safer spot to complete his "dismounting" while the rig's former owner is left to face repair bills to his car that may excede the value of the radio he was trying to protect.

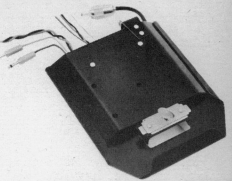

Locking type slide mounts like this one from Metro Sound are a deterrent to hit-and-run thieves. Be sure to pick a well-built, sturdy model with a good lock and adequate wiring for both power and antenna.

A better bet for locking type mounts are those that mount on the transmission hump. These are secured with heavy bolts that go right through the car floor and can be removed only from under the car—no project for a hit-and-run thief. Whether under-dash or hump mounting, however, the thing to remember when choosing a locking type mount is that it'll be no more secure than its lock. A mount made of boiler plate secured with hardened steel bolts is still useless if its security depends on a flimsy, easily-picked lock.

Still another approach to making a secure mount is a recently announced "safe" for CB or other compact electronic gear. It's a rugged steel box which bolts to the transmission hump or other convenient spot on the floor. Necessary cabling runs through a hole in the rear, and the radio itself mounts in the lid. When in use the lid is propped open, positioning the radio at a convenient angle, and for storage the lid is closed and fastened with a very heavy padlock.

TAKING IT ALONG

A safer though less convenient method of protecting your radio is to take it with you when you leave your car unprotected. This is a real nuisance if you use the standard mounting bracket that comes with most radios. Those "easy-to-remove" thumbscrews aren't all that easy if you're taking your radio out of the car every time you park it. A better bet is a slide-in mount of the type originally developed for stereo tape decks. Heavy-duty models, some with fairly secure locks for the times you don't want to bother removing the radio, have been developed specially for CB use. These also have the advantage that power and antenna connections are made automatically when the slide-in portion of the mount is seated.

This traveling businessman has a complete CB station in his attache case, ready to put into a rented car at his destination. The Antenna Specialists' magnet mount travels broken down for storage, and power for his Johnson Messenger comes from a cigarette lighter plug. If he's smart, he'll also take it with him whenever he leaves the car, even for a few minutes.

94 Security

Better protection for the radio is provided by a "custom" carrying case, tailored for the individual set. The foam interior panels in this Falcon Enterprises "CB-Saver" can be cut to provide secure storage for both radio and accessories.

Greater user convenience is provided by a portable package such as this one by Kris. With front panel, speaker grill and antenna accessible without opening the case, it need only be plugged into a power source to go on the air.

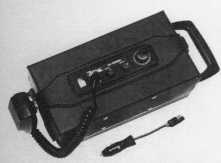

Transcom assembles different manufacturers' equipment into a variety of portable packages. This one includes two large speakers and an automatic polarity inverter for positive ground cars with a Zodiac M-5023 CB set.

This Transcom Model 2701 has an XTAL CB transceiver plus an AM/FM tuner with 8-track tape deck and three speakers.

Another approach used by a number of the "if I take it a thief can't" school is simply to not mount the rig in the car at all. Small radios will usually fit on top of the dash (fastening the factory supplied mounting bracket under the radio lets you hear the speaker audio on those models with speakers on the bottom), and larger sets can be laid on the seat or even propped up on the front edge of the seat cushion with their rear on the hump. Catering to this school of thought as well as to the group who want to operate from rental or company fleet cars, several manufacturers now offer specially designed carrying cases for CB radios or even complete prepackaged portable stations. One factor can't be overlooked when you're thinking of going the unmounted route, however. Your automobile comprehensive insurance (assuming your company isn't one of those that has already stopped covering add-on electronic equipment) *does not* cover accessories that are not mounted in the car. On the other hand, if you're a home owner, you are probably covered for "unscheduled personal property away from the premises" so—literally—all is not lost! At least one manufacturer (Johnson) offers a CB Theft Protection Plan that guarantees replacement if the set is stolen.

Dragging your CB set with you everywhere you go can be a problem, and a fair number of CB rigs have been ripped off while sitting in a "secure" place waiting for their owners to get ready to return to their cars. One way around this problem is simply to stow the rig someplace in the car such as the trunk where it's out of sight. Trunks do get broken into, of course, but not nearly as often as CBs get ripped off. You don't even need to go to the trunk to hide the radio. Simply sliding it under the seat (if your antenna and power cables are long enough you can even leave them attached) will discourage a thief looking for an easy mark. Some of the super compact models should even fit into your car's glove compartment. Some CBers have even run their set's cabling in through the back of the glove compartment so it would be ready to use when the glove compartment door was opened (inconvenient!) or could be pulled out to lay on the seat or top of the dashboard for use (long wires, again). In either event, if it can't be seen, it's a lot less likely to be stolen—for the CBer who wants to hang on to his rig, advertising definitely doesn't pay!

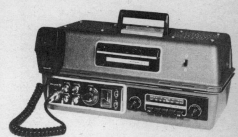

Another Transcom combines a Royce CB transceiver and separate Motorola AM/FM receiver and 8-track tape deck.

ANTENNAS

Your antenna is, unfortunately, the most prominent advertisement to the presence of a valuable piece of easily salable radio gear in your car. Looking back to the discussion of mobile antennas in the antenna section, you might almost say the better the antenna, the louder it shouts "steal me" to the prowling rip-off artist. Conversely, if your automobile gives the would-be thief no external clue that it contains two-way radio gear, your chances of hanging onto that gear are going to be greatly improved.

There are several ways to stop your antenna from being a CB thief's number one tip-off. First and most obvious is to use one of the previously discussed temporary antennas—a magnet mount or gutter clip type—and remove it whenever you park. This method has its drawbacks, particularly with respect to running the feedline in and out of a door or window each time you leave the car or return.

A better answer is a disguise antenna, one that looks like a conventional broadcast antenna but is tuned to work at 27 MHz. Some disguise antennas are also designed to operate simultaneously as broadcast antennas with no switchover needed. Before deciding on a disguise antenna be sure, if you plan a front fender installation, you can in-

A rolling "antenna farm" like this one is an engraved invitation to a CB thief.

Conventional broadcast antennas can also be used on CB with the aid of a coupler such as this Lake Model 210. To operate properly, the coupler must be adjusted with the aid of an SWR meter and the antenna always set to that length when the CB is to be used.

Antenna Specialists' MR 264 looks like any ordinary broadcast whip though it's actually a dual purpose CB/BC antenna. It utilizes a special coupler that enables it to function on both AM and FM broadcast reception at the same time it's being used on CB.

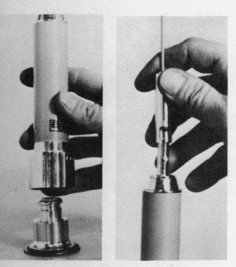

A low profile mount is a less obvious invitation to a would-be thief than a complete antenna waving in the breeze. Antenna Specialists' M-460 has a keyed arrangement to permit only the owner to remove it when parking. The M-450 is a quick-release kit for instant whip removal.

deed install it. It's a major operation to get inside the front fender of most current model cars. Several manufacturers also offer antenna adapters or tuners for mobile installations. These are ingenious little circuits that let the regular broadcast antenna function as a fairly efficient 27 MHz antenna as well as for broadcast reception.

An interesting variation of the disguise antenna is the concealed antenna—a motor-driven antenna that can be wired up to your set so it extends automatically when the set is turned on and retracts when it's turned off. They are expensive but should be a little more effective CB antenna than most disguise antennas because they are longer and lend themselves to a better location on the car body.

The third possibility for keeping your antenna from attracting unwanted attention is one of several types of quick dismount trunk lip mounting adapters. Depending on the model these adapters permit the antenna to be swung down into the trunk or removed entirely for stowing in the trunk. They are single lever released and designed to be used with any conventional trunk lip antenna. Since the trunk lip antenna is one of the best compromises of the various CB antennas, this is a good way to go—particularly if you're one who prefers to stow the rig in the trunk whenever it's out of service.

OTHER ADVERTISING

Needless to say, if the presence of a CB antenna on your car is a bad idea for security reasons, any other clue that you've got CB is equally bad. In fact, windshield or bumper stickers proclaiming the car owner's enthusiasm for CB may even make him a more likely target! It's the dedicated CBer who is more likely to have the more expensive, multi-featured radio under his dash along with antenna meters, power mikes and other easy-to-fence goodies—and whose rear window proclaims "We Monitor Channel 9," "Toot twice if you're into CB," and the like. A car so marked promises a lot better haul than one with a short inconspicuous antenna on the rear deck. If you must advertise your CB enthusiasm for jamborees, club activities or emergency work, do it with some sort of removable sign you can stick in your window when needed but can remove when you are done. Day-to-day advertising is asking to be ripped off!

BURGLAR ALARMS

In this day and age a burglar alarm for your car is almost a must, whether or not you're running a CB set in the car. Automobile "hit men" have become so accomplished that they'll open up a locked car to steal a shopping bag or briefcase off the seat or even go after the unknown contents of the glove compartment. Along with the current escalation in car break-ins, however, there has also developed a very competitive market in auto burglar alarms. Most are easy to install, reasonably priced and fairly effective.

Auto alarm systems fall into two major categories, the so-called "hard-wired" type and the electronic systems. Their names are derived from how they are set off. A hard-wired system is triggered by the opening or closing of a switch tied directly to the alarm. It is the least expensive of the two types since its circuit is very simple. It's also usually considered more reliable since it has less to go wrong with it, but it's also more trouble—a *lot* more trouble—to install. Its problem is that every opening you wish to protect, trunk and hood as well as all doors, requires a switch and wires running from each switch to the alarm.

Electronic alarms are much simpler to install because the car's electrical system provides most of the built-in wiring they need. They are triggered by any rapid minute change in the voltage supplied by the car battery. Such a change occurs when any electrical device in the car is turned on. Thus the alarm will go off when the car door opens—even the dome light will set it off. Trunk and hood lights will also set it off if your car has them as factory supplied items, and they can easily be added if it doesn't.

Some alarm systems feature a built-in delay so they can be set from inside the car but still give the driver time to get out before they are armed and ready to go off. Conversely, they have a built-in delay between the time they are triggered and the time they go off to permit the driver to get back in his car and deactivate them. Though this seems like a good idea in theory, it has not always worked out in practice. Some CB thieves have become so proficient they've been able to get into a delay-type alarm equipped car and have the radio halfway removed before the alarm went off. In at least one reported case the gutsy thief just kept on with what he'd started and was out of the car, walking away with the radio under his coat, before anyone responded.

The instant-on alarms require some sort of external switch, almost always in the form of a key-operated lock switch, to turn them on and off. These switches are

The spring-loaded latch of this lip-mounted antenna permits its release in an instant for trunk storage while the user's car is parked. (South Com)

This folding mount is bolted to the side of the trunk opening, permits the antenna to be swung into the trunk for concealment. (Shur-Lok)

This ingenious alarm, fastened to the CB set (or other accessory to be protected) with a Band-Aid like tag, sounds the car horn if the tag or connecting wire is cut loose or torn. (Mountain West Alarm)

mounted in some convenient spot, usually on the left front fender, and are a dead giveaway that the car is alarm equipped. This isn't necessarily a drawback, of course. A thief who has a choice between breaking into a car equipped with an alarm and one without would be wise to tackle the latter. However, there are ways to defeat these types of alarms—which we won't go into here—so making your alarm system obvious isn't always an advantage. Some cars offer a neat opportunity to hide your alarm lock, by the way, among the numbers, letters or just plain decorative trim that often adorns front fenders.

There are also some burglar alarms designed to protect the rig only. One type is hooked into the antenna feedline, and if it's cut or the connector on the rear of the rig is disconnected, it goes off. Such an alarm has one big advantage—it's left on all the time so doesn't require any thought except when you want to remove your radio. The CB-only alarm has one big disadvantage—it doesn't protect anything else in your car, and with tape decks, stereo speakers, packages, cameras and the car itself also prime rip-off artists' targets, it would seem wiser to invest your time and money in a more complete form of protection.

The least expensive alarm systems are those that operate your automobile horn. These have two disadvantages, the first that a blowing car horn doesn't usually command much attention and the second that a car horn draws enough current that it could run your battery down if it ran long enough. Alarm systems that include the alarm use either mechanical or electronic sirens and thus sound like emergency vehicles. As a result, one of them is much more likely to attract passerby interest and—conversely—scare off the would-be thief.

If you do decide to put in a burglar alarm, do shop around. There's a lot of variation in quality, ease of installation and cost. Don't delude yourself into thinking your new alarm system will provide you complete rip-off protection. You'll be appalled to learn how little attention a wailing siren actually receives those first few times you forget about it and set it off in the supermarket parking lot. Do test it from time to time because wires break, connections corrode and moisture gets into delicate electronic circuits. An alarm system that can't be depended on is worse than none at all!

OTHER PROTECTION

Check your insurance. Discuss your automobile comprehensive, motor club (if any) and home owner's policies with the appropriate representative. Find out where and for how much you're covered in each policy. If it's not clear or the representative promises something you don't find in the policy, ask him to confirm it in writing and put his letter with your policy. You may have to pay an added premium to cover your radio, and if so, be careful of any deductibles. It would not be good economics to pay $20 a year to protect a $109.95 radio when you have a $100 deductible policy!

Do everything you can to facilitate recovery of your radio if it is stolen. Report it promptly to the police and provide them with full details on the spot. Carry a note with make, model and serial number with your driver's license so you'll be ready if and when that horrible event happens to you. Get the name and badge number of the officer who took your report along with his district or department and the complaint number. All may come in handy if there is a later recovery or a hassle over insurance.

Send in the warranty registration card that came with your radio. That sounds obvious, but it isn't. Apparently many people keep the cards to send in after the radio fails in the belief that no matter how long they've had the radio they can lie and claim it's still under warranty. The manufacturers rarely fall for this. They're not dumb enough to think a radio they shipped into the hot CB marketplace 2 years ago just got sold last week, and many of them do maintain files of all their warranty cards by serial number. If your stolen radio ever came back for service or is recovered by police, the warranty card just could be your key to its return.

One fault common to almost all CB radios is their lack of permanent identification. The nameplate which bears the radio's serial number is almost invariably a stick-on or screw-on device, readily removed by thieves to prevent tracing. Though some states have laws against selling merchandise without serial numbers, these laws are spottily enforced at best.

The problem of removable serial numbers is going to be helped by new rules announced by the FCC this past summer which go into effect January 1, 1977. However, these rules apply only to new model radios that come on the market after that date, so there are literally tens of millions of CB sets those rules don't cover. You can help protect your radio and facilitate its return to you if it's ripped off by engraving the serial number and your own driver's license number somewhere on the chassis with an electric

"Do everything you can to facilitate recovery of your radio if it is stolen. Report it promptly to the police and provide them with full details on the spot."

100 Security

engraving tool. The serial number, marked as "S/N 1001234" positively identifies the radio (you *did* keep your bill of sale, didn't you?) while your driver's license number, for example "IL S636-0001-1234" enables it to be quickly traced back to you by computer. It's amazing how many recovered radios, even with serial numbers intact, police are unable to return to their owners because of insufficient information.

One last tip that's paid off in a few rip-off cases. Make some non-obvious identifying mark on the front panel of your radio—a little black paint in the corner of a white-painted letter, or an extra dimple or scratch in a bezel design, for example. Such an "ID" is obvious to your eye and no one else's, and several ripped off owners have actually spotted their missing sets in flea markets and—in one case—under the dash of an acquaintance's car!

DIRTY TRICKS

In addition to all the ideas to help preserve your radio that have been presented in the preceding pages, there are

"Make some non-obvious identifying mark on the front panel of your radio. Such an "ID" is obvious to your eye and no one else's."

some other little tricks that have been used by various CBers to slow down or prevent rip-offs. They may or may not accomplish what they're supposed to—some may not even be legal in every state—but they're offered here purely as information, not as recommendations.

One way to confuse a would-be thief is to let him think you have a radio he doesn't want. Some CBers have done this by picking up a control box for an old, tube-type police or taxi radio and mounting it under the dash. Such radios have little current value—none on the CB market—yet the presence of one in a car could explain why that car has a two-way radio antenna. The CB rig would, of course, be tucked away in the glove compartment or under the seat so it's out of sight. Another way to discourage a would-be thief is to put warning decals for burglar alarms and even put a key switch in the front fender. You don't need the alarm itself, the argument goes, just the threat of one will scare off most thieves.

Under the heading of dirtier tricks comes various means of booby trapping the radio installation itself. One device available commercially for this purpose is a tear gas cartridge. It's mounted up under the dash, and a string is tied from its ring trigger to a convenient point on the rear of the rig. When unknowing hands fumble around behind the radio or attempt to remove it, bang!—a face full of tear gas is the result. It does work. One ham operator, investigating his car after hearing a disturbance in his driveway late one night, found his car door half open, his mobile rig half out of its mount—and blood on the floor of the car.

A much dirtier trick has been used by a few nastier-minded CBers. Their philosophy is to let pain be a deterrent, so they've taped razor blades or fish hooks along the rear edge of their radios where thieving fingers would be likely to grasp. The argument against such a drastic approach should be obvious, considering that you, your family, friends and service station people all are likely to stick fingers or hands up behind the dash on legitimate business! Just remember that your set's booby trapped when you go to remove it yourself.

If it seems that the security problem has received a lot of emphasis in these pages, it's because it deserves it. Rip-offs of CB radios have become a major crime problem, yet there *are* a number of things that have been discussed here that could help you to avoid becoming a crime statistic. We hope you use them—and you *do* keep your rig!

chapter 9

CB LINGO

how can you communicate when you can't speak the language?

Understanding what's coming over a CB radio is one of the toughest tasks faced by a newcomer to CB. Under the best of conditions two-way radio communication sounds clipped and mechanical, and add to that a little interference, some technical terms, a dash of CBese that's still a long way from appearing in any collegiate dictionary, serve the whole mess up in a high-speed sing-song and nine out of 10 non-CBers will swear they're listening to a conversation in Swahili.

Probably the best starting point to understanding what you hear when you turn on your CB set is to break the CBer's unique language into its prime components and explain what each is and where it came from. For purposes of this discussion, that breakdown can consist of 4 parts: 10 code, Q signals, technical language, and "CBese." Since the first two can be most mysterious to listeners unfamiliar with them, we'll start there.

10 CODES

As anyone who watches television knows, the 10 code is used by police to shorten and—to some degree—disguise the meaning of their radio communications. It developed in

(Facing page.) For the businessman who finds himself on the road much of the time a CB can provide timesaving road directions or road conditions information. (Johnson)

the very early days of police radio when the police used broadcast stations to notify their cruisers of crimes in progress, a practice that had an obvious disadvantage— major crimes or disasters would bring such crowds of curious listeners that responding police would have a hard time reaching the scene! Over the years the 10 code grew and saw many changes, reaching over 100 signals in some versions. It not only became very complicated, but local and regional variations made it very confusing (10-10, for example, meant "standing by" to some police departments but "fight in progress" to others!).

With neighboring police forces working more closely together this type of confusion became intolerable so in 1975 APCO, the Associated Public Service Communications Officers, issued a greatly simplified 10 code consisting of only 34 signals. Many but not all of the signals in APCO's new 10 code are the same as they were before, but some such as 10-10 (which now means "negative") have been changed. Though a lot of police forces around the country have adopted the new code, many are still using the old—eventually the new code should become universal.

The 10 code is being used extensively by CBers, but the code they're using is the old one. The version included here is one of the most common versions—with it you should be able to decipher about any 10 signal you'll ever hear on 11 meters. For the benefit of those who monitor the public service bands, the new APCO 10 code is also included. You'll have to figure out whether your department is using the old or new code by listening.

OLD 10 CODE

10-1	Signal Poor	10-18	Complete Assignment Quickly
10-2	Signal Good	10-19	Return to Base
10-3	Stop Transmitting	10-20	Location
10-4	Acknowledgment	10-21	Call By Phone
10-5	Relay	10-22	Disregard
10-6	Busy, stand by	10-23	Arrived at Scene
10-7	Out of Service	10-24	Assignment Completed
10-8	In Service	10-25	Report in Person
10-9	Repeat Transmission	10-26	Detaining Subject
10-10	Fight in Progress (Standing by)	10-27	Drivers License Info.
10-11	Dog Case	10-28	Vehicle Registration Info.
10-12	Stand by, Stop (Visitors Present)	10-29	Check Records
10-13	Weather and Road Report	10-30	Illegal Use of Radio
10-14	Report of Prowler	10-31	Crime in Progress
10-15	Civil Distrubance	10-32	Man With a gun
10-16	Domestic Trouble	10-33	Emergency
10-17	Meet Complainant	10-34	Riot

Code	Meaning	Code	Meaning
10-35	Major Crime Alert	10-69	Message Received
10-36	Correct Time	10-70	Fire Alarm
10-37	Investigate Suspicious Vehicle	10-71	Advise Nature of Fire (size, type, etc.)
10-38	Stopping Suspicious Vehicle		
10-39	Urgent—Use Red Light & Siren	10-72	Report Progress of Fire
10-40	Silent Run—No Light or Siren	10-73	Smoke Report
10-41	Beginning Tour of Duty	10-74	Negative
10-42	Ending Tour of Duty	10-75	In Contact With
10-43	Information	10-76	En Route
10-44	Request Permission to Leave	10-77	Estimated Time of Arrival
10-45	Animal Carcass	10-78	Need Assistance
10-46	Assist Motorist	10-79	Notify Coroner
10-47	Emergency Road Repairs Needed	10-80	Chase in Progress
10-48	Traffic Standard Needs Repair	10-81	Breathalizer Report
10-49	Traffic Light Out	10-82	Reserve Lodging
10-50	Accident	10-83	School Crossing
10-51	Wrecker Needed	10-84	Advise Estimated Time of Arrival
10-52	Ambulance Needed	10-85	Delayed
10-53	Road Blocked	10-86	Officer/Operator on Duty
10-54	Livestock on Highway	10-87	Pick Up Checks for Distribution
10-55	Intoxicated Driver	10-88	Advise Present Telephone Number
10-56	Intoxicated Pedestrian		
10-57	Hit & Run	10-89	Bomb Threat
10-58	Direct Traffic	10-90	Bank Alarm
10-59	Convoy or Escort	10-91	Pick Up Prisoner/Subject
10-60	Squad in Vicinity	10-92	Improperly Parked Vehicle
10-61	Personnel in Area	10-93	Blockade
10-62	Reply to Message	10-94	Drag Racing
10-63	Prepare to Make Written Copy	10-95	Prisoner/Subject in Custody
10-64	Message for Local Delivery	10-96	Mental Subject
10-65	Net Message Assignment	10-97	Check (Test) Signal
10-66	Message Cancellation	10-98	Prison or Jail Break
10-67	Clear to Read Net Message	10-99	Records Indicated Wanted or Stolen
10-68	Dispatch Information		

NEW 10 CODE

Code	Meaning	Code	Meaning
10-1	Signal Weak	10-20	Location
10-2	Signal Good	10-21	Call (name) by Phone
10-3	Stop Transmitting	10-22	Disregard
10-4	Affirmative (OK)	10-23	Arrived at Scene
10-5	Relay (To)	10-24	Assignment Completed
10-6	Busy	10-25	Report to (Meet)
10-7	Out of Service	10-26	Estimated Arrival Time
10-8	In Service	10-27	License/Permit Information
10-9	Say Again	10-28	Ownership Information
10-10	Negative	10-29	Records Check
10-11	(Name) On Duty	10-30	Danger/Caution
10-12	Stand By (Stop)	10-31	Pick up
10-13	Existing Conditions	10-32	( ) Units Needed (Specify: Number/Type)
10-14	Message/Information		
10-15	Message Delivered	10-33	Help Me Quick
10-16	Reply to Message	10-34	Time
10-17	Enroute		
10-18	Urgent		
10-19	(In) Contact		

Q SIGNALS

Like 10 codes, Q signals have been adopted from other services by the CB clan. Q signals also have an ancient and honorable history, having developed in the early years of radio communications as a rapid means of asking questions or answering them. To the CW* (code) operator "QTH?" for example, asks "What is your location?" while "QTH San Francisco" supplies the answer, "My location is San Francisco."

Amateur radio operators have always used the Q code extensively, and though it was developed as a tool for CW operators, it is also widely used by hams operating on voice. With a number of hams also active as CB operators and many CBers also shortwave listeners who monitor the ham bands, it's only natural that the most often used ham Q signals also show up on 11 meters. Here are the ones you're most likely to hear used by CBers:

Q SIGNAL MEANINGS

QRM	Interference from other stations	QSL	Confirmation of contact, usually by postcard
QRN	Noise interference (static), natural or man-made	QSO	Contact or communication
QRT	Stop operating, shut down	QSY	Move to a different channel or frequency
QRX	Wait, stand by		
QSB	Signal fading	QTH	Location

TECHNICAL LANGUAGE

The CB operator's technical language consists primarily of those terms having to do with performance and operation of his rig and antenna system. Many of these terms are familiar to the hi-fi enthusiast, and almost all are in the everyday vocabulary of a radio amateur or two-way radio professional. Only a few are unique to CBers.

One word of caution when you are dealing with some CB operators on technical matters. Like most hobbyists, CBers represent a very broad cross section of the population and bring a wide variation in technical competence to the 27 MHz band. Some CBers are graduate electronic engineers who've participated in the design of the equipment they're using, while the technical abilities of others barely qualify them to successfully plug a base station power plug into a wall socket. All too often, unfortunately, it's the latter breed who can be found speaking with great conviction on the solution of knotty technical problems. Beware!

---

*CW—"Continuous Wave" (Morse Code)

TECHNICAL TERM	DEFINITION
AC	Alternating Current, the type of power found in most homes.
AGC (or AVC)	Automatic Gain (Volume) Control. A receiver circuit that keeps loudspeaker volume constant on stations of different strengths.
AM	Amplitude Modulation. A system for modulating a radio transmitter's carrier by varying its strength.
Amperes (Amps)	Unit of electrical current.
ANL	Automatic Noise Limiter. A circuit used in a receiver to lower interference from a car ignition and other electrical noise sources.
Band	A group of frequencies or channels. Class D CB is in the 27 MHz band.
Bandwidth	The amount of spectrum taken up by a signal.
Base Station	A station at a fixed location operating from house power lines.
Beam	A directional antenna.
Carrier	The radio wave that carries the modulation.
Channel	A specific spot or frequency.
Clarifier	A control found on single sideband sets to adjust receiver tuning and make the incoming signal clearer.
Class D	The most popular CB license for communicating on 27 MHz. The other CB band is Class A. Class C CB uses special 27 MHz channels for controlling models by radio.
Coax, Coaxial Cable	Pronounced "co-ax," the type of shielded cable used to connect a transmitter to the antenna.
Crystal	Frequency determining device that sets a receiver and/or transmitter to an exact frequency or channel.
DC	Direct Current, the type of power delivered by a battery. Vehicle voltage is usually 12 volts.
dB (decibel)	Comparison between two power levels. Used to compare antennas and receiver sensitivity, etc. For reference, a 3dB increase is twice as much power, and 10dB is a 10 times increase.
Delta Tune	Circuit permitting fine-tuning stations transmitting slightly off frequency.
Dual (or double) Conversion	A type of receiver circuit with superior interference rejection.
Dummy Load	A simulated antenna used for testing transmitters without causing on-the-air interference.
11 Meters	Wave length of the 27 MHz band.
FET	Field Effect Transistor.
Field Strength Meter	Instrument that measures the strength of a radiated signal.
Frequency Synthesis	Method used to reduce number of crystals in a multi-channel radio.
Fuse Block	A fixture that holds a group of fuses mounted behind the dash of an automobile.
Gain	An increase in signal strength or power.
Ground	Connection to earth or a common wiring point.
Ground Plane	Type of nondirectional antenna for base station use.
Handset	Combined microphone and earphone, like a telephone.
Handie-Talkie (HT)	Radio designed for portable operation with a built-in antenna and batteries for power.
IC	Integrated Circuit. Tiny semiconductor device containing many circuit elements.
Impedance	Amount of opposition a circuit exhibits to a flow of current. Antennas, feedlines and transmitter output circuits all have a "characteristic impedance" which must be matched for highest efficiency.
Input power	Power delivered to the output stage of a transmitter. Class D CB is limited to 5 watts input.
Ionosphere	Stratospheric layer that reflects radio waves for long distance communication.
Jack	Socket for quick connection of a microphone or speaker.
kHz	Kilohertz. Term meaning 1000 cycles per second.
LED	Light-Emitting Diode. A type of electronic indicator or lamp.
Linear	Amplifier to boost transmitter power. Illegal on 27 MHz CB.
Loaded Antenna	Antenna which has been shortened by use of a loading coil.
Loading coil	Coil used to electrically shorten an antenna.
MHz	Megahertz. Term used when designating frequency of a radio signal. Mega is one million and Hertz (named after a radio pioneer) is cycles per second. CB is on the 27 MHz band, meaning a CB radio wave oscillates about 27 million times per second.
Microvolt	One millionth of one volt.
Mike	Microphone.
Mobile Rig	Transceiver operated in a car or other vehicle.

108 CB Lingo

TECHNICAL TERM	DEFINITION
Modulating	Causing the carrier, or radio wave, to vary with voice or music.
Noise Blanker	Type of noise reducing circuit.
Ohm	Unit of electrical resistance.
Omnidirectional	All directional or nondirectional.
Oscillator	An amplifier connected so that some of its output is fed back into its input and reamplified, generating an audio or radio frequency signal.
Output Power	Number of watts delivered by a transmitter. For class D CB the limit is 4 watts.
PEP	Peak Envelope Power. Method for rating the power of a single sideband (SSB) transmitter.
Phonetic Alphabet	Words used in place of letters to reduce errors in voice communications.
PLL	Phase Lock Loop. Type of synthesizer circuit.
PTT	Push-To-Talk
Quarter Wave Antenna	Shortest basic antenna.
RF	Radio Frequency.
Rig	Transmitter or transceiver.
RFI	Radio Frequency Interference.
Selectivity	Ability of a receiver to reject off-frequency stations.
Sensitivity	Ability of a receiver to detect and amplify weak, distant stations.
Shielded Cable	Cable with an inner wire surrounded by an outer metal braid or shield.
Signal	Radio signal.
Signal Strength	Strength of incoming signal, usually measured on S-meter.

TECHNICAL TERM	DEFINITION
Skip	Long-distance transmission of radio signals by reflection from the ionosphere.
S-Meter	Meter used in a receiver to measure strength of incoming signal.
Squelch	Circuit that silences the receiver when there is no incoming signal.
Squelch Control	Control that adjusts a receiver squelch circuit to eliminate background noise in the loudspeaker when no signal is being received.
SSB	Single Sideband. A method of transmission in which the carrier and one sideband are eliminated from the transmitted signal, reducing signal bandwidth and increasing efficiency.
SWR	Standing Wave Ratio. An indication of how well an antenna system is tuned.
Synthesizer	Type of frequency generating circuit that develops many frequencies from one or two crystals.
Traffic	Messages
Transceiver	A combination transmitter and receiver.
Transmitter	Equipment capable of sending but not receiving.
TVI	Television Interference from a transmitter.
Volt	Unit of electrical pressure.
VOX	Voice-Operated circuit to automatically turn on a transmitter when the microphone is spoken into.
Watt	Unit of electrical power.

Moving barges down a narrow canal can be hazardous. With the help of two-way radio a coordinated effort can be faciliitated.

PHONETIC ALPHABET

Many letters sound almost the same when spoken, so two-way radio operators often substitute words for letters to reduce the chance for error due to interference or weak signals. Though humorous, geographical or other forms of phonetics are often used (KLF 1234 might identify as "Kissing Lovely Females 1234" or "Kansas, Louisiana, Florida 1234" for example) the International Phonetic Alphabet should always be used when handling emergency or other urgent traffic. It is:

International Civil Aviation Organization (ICAO) Phonetic Alphabet

A - ALFA	H - HOTEL	O - OSCAR	V - VICTOR
B - BRAVO	I - INDIA	P - PAPA	W - WHISKEY
C - CHARLIE	J - JULIETT	Q - QUEBEC	X - X-RAY
D - DELTA	K - KILO	R - ROMEO	Y - YANKEE
E - ECHO	L - LIMA	S - SIERRA	Z - ZULU
F - FOXTROT	M - MIKE	T - TANGO	
G - GOLF	N - NOVEMBER	U - UNIFORM	

Example: W1AW . . . W 1 ALFA WHISKEY . . . W1AW

This trucker has added character to his 18-wheeler. Note his wife's name on the front of the truck and his call sign and name on the door of the cab.

"HANDLES"

CB handles provide an interesting perspective on CB communications. They seem to have come into existence primarily as a result of the long-standing battle between the FCC and the CBers not willing to go along with some of the Commission's Rules. The rules restricting contact length and subject were two of the prime culprits, so CBers who wanted to ragchew about their equipment "went underground" by not using call letters or even proper names to make themselves harder to identify by FCC monitors.

The Commission's relaxation of some of their restrictive rules in the fall of 1975 did away with much of the "necessity" for using handles, and proper names (just like call signs) are heard on 11 meters much more today than they used to be. However, handles have a value and purpose even when CBers are 100 percent within the rules—there are a limited number of proper names in the U.S. and all but a very few are shared by a multitude of others.

"Axehandle," "Partyboy" and "Sweet Thing" are a different story, however. It's not all that tough to come up with a handle you've never heard before and for your trouble your 11 meter identity will have a lot more character than any FCC assigned call letters could ever provide.

For the small boater CB is many times the only way of communicating with other boaters or with friends and family on land.

Short range hand-held CBs help this man direct the loading of this boat. (Standard)

That's why signoffs like "This is KAB 5678, The Blue Baron, on the side!" are quite acceptable these days.

CB TERMS

Although CB operators speak English, for the most part, they have developed an extensive vocabulary of words unique to their service. Some of these like "Smokey" have become well-known from the publicity CB radio has received, but many are obscure enough that the newcomer to CB will find them next to impossible to figure out.

"CBese" varies from one part of the country to another, and sometimes even between local clubs. A "complete" dictionary of CB terms would make a book in itself and, in fact, such a CB dictionary has been compiled. However, like a tourist visiting a friendly foreign country, a beginning CBer with a good basic vocabulary will find he can not only understand most of what the natives are saying but will be able to communicate with them too! The CB lingo that appears below is such a basic vocabulary, so try it and see.

CB JARGON

ACE—An important CBer
ADVERTISING—A marked police car with its lights flashing.
AFFIRMATIVE—Yes
ALL THE GOOD NUMBERS ON YOU—Wish you the best!
ALLIGATOR—All mouth and no ears; usually a station with a good signal but poor receiver.
ANCHORED MODULATOR—Base station
APPLE—A CB addict
APPLIANCE OPERATOR—An operator who doesn't know anything about his radio.
BACK DOOR—Last vehicle in a string of three or more in contact with each other by CB.
BACK HER DOWN—Slow down
BACK IN A SHORT—Be right back
BACK OFF—Slow down
BACK OUT—End contact
BAD SCENE—A crowded channel
BALLET DANCER—An antenna that really sways
BAREFOOT—Using an unmodified CB transmitter; without linear amplifier.
BARN—Truckers' terminal
BASE, BASE RIG—Station at home, office, or other fixed location
BASEMENT—Channel 1
BEAM—A directional antenna
BEAN STORE—Restaurant or road stop where food is served
BEAR—Policeman
BEAR BAIT—Motorist without CB; fast driver
BEAR CAVE—Police station or post on highway
BEAR IN THE AIR—Police in helicopter or airplane
BEAR NEST OR DEN—Police headquarters or state patrol barracks
BEAST—A CB rig
BEAT THE BUSHES—"Front door" (Lead vehicle) looks for Smokey.
BEAVER—Gal, other half, female CBer.
BE-BOP—Radio Control signals
BEER TONE—An intermittent tone signal
BIG DADDY—Federal Communications Commission
BIG EARS—Good receiver
BIG 4-10—Emphatically yes (10-4)
BIG SWITCH—On-off knob on CB set
BLEED-OVER—Interference from station on an adjacent channel
BLESSED EVENT—A new CB rig
BLOOP BOX—Ambulance
BOAST TOASTIE—A CB expert
BOAT ANCHOR—A big old radio
BOBTAIL 18 WHEELER—Tractor only, no trailer

BODACIOUS—Loud, sounds good
BOOSTER—Amplifier
BOOTLEGGER—Operator without a license, or who is using a false call sign.
BOOTS, SHOES, GALOSHES—Linear amplifier
BOULEVARD—Highway, interstate
BOY SCOUTS—The State Police
BREAKER—Request to break-in on a channel (break or breaker is repeated followed by the channel number)
BREAKING UP—Signal is cutting on and off.
BREAK ONE-OH—Also "Break 10"—I want to talk (on channel 10).
BRING 'ER DOWN—Slow down
BRING IT ON, BRING YOURSELF ON, BROUGHT IT ON—Go ahead, it's clear
BROWN BOTTLES—Beer
BTO—Big Time Operator
BUBBLE-GUM MACHINE GOING—Lights flashing on police car.
BUCKET MOUTH—Idle conversation
BUG OUT—To leave a channel
BURNING UP MY EARS—Got a good signal
BUSHELS—One-half ton; a 20-ton load would be 40 bushels.
CALL YOU IN A SHORT—Be right back
CAMERA—Police radar
CAN—Shell of a CB set
CANDLE LIT—Lights on or lights flashing on police car
CANDY MAN—FCC man
CARTEL—A group hogging a channel
CASH REGISTER—Toll booth
CATCH A FEW "Z'S"—Get some sleep
CATCH YOU ON THE OLD FLIP/FLOP—Talk to you on the return trip
CELL BLOCK—Location of the base station
CHAIN GANG—Members of a CB club
CHANNEL 25—The telephone
CHARLIE—The FCC
CHECK THE SEATCOVERS—Watch out for a car with a pretty female driver or passenger
CHICKEN COOP—Highway truck weigh station
CHOPPED TOP—A short antenna
CHROME DOME—Mobile unit with a roof antenna
CLEAN—No police in sight
CLEAN CUT—An unmodified rig
CLEAN SHOT—Road is clear of Smokeys, good road conditions
CLEAR—Message complete
COFFEE BREAK—Social meeting of local CBers; an "unorganized" CB social get-together, as opposed to a planned out Jamboree. Often held at a diner.

COME BACK—Repeat message
COME ON—Over to you; answer me
COMIC BOOKS—Truck drivers' log sheets or log books
CONTAINER—Chassis and shell of a CB rig
COPY—Talk, listen, any kind of radio contact
COPYING THE MAIL—Listening to channel chatter
COTTON PICKER—Term used instead of four-letter words on the air
COUNTRY CADILLAC—Pickup truck
COUNTY MOUNTIE—County sheriff or highway patrol
COVERED UP—Blanked out by another station
CRADLE BABY—CBer who is afraid to ask someone to stand by
CUB SCOUTS—Sheriff's men
CURLY LOCKS—Coils in a CB rig
CUT OUT—To leave a channel
DADDY-O—The FCC
DARKTIME—Night
DEFINITELY—Maybe
DESPAIR BOX—Box where spare CB components are kept
DOG BISCUITS—dB, decibels
DO IT—TO IT—Put the hammer down and go
DON'T TENSE—Take it away
DOUBLE-BARREL PICTURE TAKER—Mobile two-way radar
DOUBLE "L"—Land line, telephone
DOUBLE NICKEL—55 MPH legal speed limit
DOUBLED—Two stations transmitted at the same time
EARS—Antennas or radios
EATUM-UP—Roadside restaurant
EIGHTY-EIGHT—Love and kisses
EIGHTEEN-WHEELER—Any semi-tractor truck with any number of wheels
EYEBALL—Look for, or already in sight; to meet face-to-face
FAT LOAD—Overload, more weight than local state law allows
FEED THE BEARS—Collect a traffic ticket
FIFTY DOLLAR LANE—Left-most lane, or passing lane
FINAL—Last transmission; final power amplifier tube or transistor
FINAL 20—Destination
FINGERS—A channel-hopping CBer
FIVE-FIVE—55 mph
FLAG WAVER—Highway worker
FLAPPERS—Ears
FLIP-FLOP—Smokey turning around; return trip
FLIP SIDE—Return trip
FOG LIFTER—Interesting CBer
FOOT-WARMER—Linear amp
FOUR—Abbreviation of 10-4, meaning OK
FOUR-TEN—10-4, Emphatically
FOUR-WHEELER—A passenger car

FOX CHARLIE CHARLIE—The FCC
FRIENDLY CANDY COMPANY—The FCC
FRONT DOOR—Lead vehicle in a group communicating by CB
FUGITIVE—CBer operating on different channel than favorite
FULL DRESS TAXI—Police car with lots of lights, markings, sirens, etc.
GEOLOGICAL SURVEY—CBer who looks under his set
GEORGIA OVERDRIVE—Neutral gear
GETTING OUT—Your transmission is clear and strong
GIVE YOU A SHOUT—Give you a call
GLORY CARD—Class D License
GO-GO GIRLS—Loads of pigs
GOING-HOME HOLE—High gear
GONE, WE GONE—Signing off, clearing channel
GOOD BUDDY—Any other CBer
GOODIES—Extra accessories for a CB rig
GOOD NUMBERS—73 and 88
GOON SQUAD—Channel hoggers
GOT YOUR EARS ON?—Are you listening?
G.P.—Ground Plane type CB antenna
GRAB BAG—Illegal hamming on CB
GRASS—Side of the road or median strip
GREASY—Slippery
GREEN STAMP—Dollar
GREEN STAMP ROAD—Tollway (or simply "Green Stamp")
GUY—A name used when another's handle is unknown
HAG FEAST—Group of female CBers on the channel
HAM—Amateur radio operator
HAMMER—Accelerator pedal
HAMMER DOWN—Highballing; driving fast
HAMSTER—One who "hams" on CB
HANDLE—Slang names used by CBers instead of call letters
HAPPY NUMBER—A S-meter reading
HENCHMEN—A group of CBers
HOLE IN THE WALL—Tunnel
HOME 20—Base station (or CBers home town)
HOUND MEN—Policemen looking for CBers using rig while mobile
HUNG UP—CBer who can't leave set
IDIOT BOX—TV set
IN A SHORT—Soon
INDIAN—Neighbor who has TVI from you
IN THE GRASS—Parked or pulled over on the median strip
IT'S ALL CLEAN—No bears
JAMBOREE—Gathering sponsored by CB club, usually consisting of social activity and equipment displays
JIMMY—GMC truck
KEEP 'EM BETWEEN THE DITCHES—Have a safe trip

KEEP THE SHINY SIDE UP AND THE GREASY SIDE DOWN—Have a safe trip
KENOSHA CADILLAC—Any car made by AMC
KEYBOARD—Controls of a CB set
KEYING—Pressing mike button
KICKER—Linear amplifier
LAND LINE—The telephone
LAY 'EM DOWN—Pull off the road
LEGALIZING—Going at the legal speed limit (55 mph)
LID—Inexperienced or poor operator
LINEAR—A power-boosting amplifier (illegal for CB)
LOCAL BEAR—City police
MARKER—Mile marker, milepost on the highway
MAKING THE TRIP—Your transmission is clear and strong
MAN IN BLUE—Policeman
MAN IN SLICKER—Fireman
MAN IN WHITE—Doctor
MAYDAY—International emergency distress call
McCLEAN LANE—Middle lane
MERCY, or MERCY SAKES—Conversational pause
MIKE—Microphone
MOBILE RIG—Transceiver installed in automobile
MODJILATING—Talking
MONFORT LANE—Left lane
MOTH BALL—Annual CB convention
NEGATORY—No, negative
O.M.—(Old Man) a CBer
ONE-EYED MONSTER—Television set
ONE-HUNDRED MILE COFFEE—Strong coffee at truck stop
ON THE MOVE—Moving or driving fast
ON THE PEG—Legal speed limit
ON THE SIDE—Parked or pulled over on the shoulder (also, standing by)
OTHER HALF—Wife (usually) or husband
OUT—Conversation is over and no reply is expected
OVER—Go ahead and speak
OVER-MODULATING—Too loud, distorted
OVER YOUR SHOULDER—Road report asked of a CBer coming toward you. "How does it look over your shoulder?"
PANIC IN THE STREETS—Area being monitored by FCC
PARKING LOT—Bumper-to-bumper traffic
PART 95—FCC rules governing Citizens Band
PART 15—FCC rules covering unlicensed, low power handie-talkies
PAVEMENT PRINCESS—Hooker at truck stop
PEANUT WHISTLE—Low powered station; also, station with no kicker
PEDAL TO THE METAL—Accelerator to the floor
PENMAN—CBer to be who has filled out FCC forms
PEPPER SHAKER—Ash spreader
PICK A CLEAN ONE—Select an unused channel

PICTURE TAKER—Police with radar
PICKUM-UP—Light truck; pickup truck
PIGGYBACK—Toll plaza on the highway
PLAIN WRAPPER—Unmarked police car
PLAY DEAD—Stand by
PORTABLE CHICKEN COOP—Portable weigh scales
PORTABLE PARKING LOT—Auto-transport trailer
PORTABLE RIG—Transceiver with self-contained antenna and batteries for use in the field
POSTHOLES, LOAD OF—An empty truck
POUNDS—S-meter reading (S-3 is three pounds, etc.)
PRESCRIPTION—FCC rules
PREGNANT ROLLER SKATE—Volkswagen
PULL SWITCHES—Shut down station
PULL THE BIG ONE—Signing off for good
PUMPKIN—Flat tire
PUT THE GOOD NUMBERS ON YOU—Best regards (for 73 and 88)
Q-BIRD—An intermittant tone generator
RAKING THE LEAVES—Back door or last vehicle in string, bringing up the rear
RANCH—Where truckers spend the night; truck stop, home, motel, etc.
RATCHET JAW—Nonstop talker
READING THE MAIL—Eavesdropping on other stations
REEFER—Refrigerated trailer
REST-UM UP—Roadside rest area
RIG—CB radio; Transceiver; Mobile rig; Portable rig, etc.
RIOT SQUAD—Neighbors who have TVI
ROCK—The crystal in a transceiver to control transmitting or receiving channels, so called because it is cut from quartz
ROCKING CHAIR—The middle vehicle between front and back doors
ROGER—O.K.
ROGER DODGER—Same as "Roger"
ROGER ROLLER SKATE—Passenger car going more than 20 mph over limit
ROLLER SKATE—Small car
SALT SHAKER—Salt spreader
SAVAGES—CBers who hog the channel
SAY AGAIN—Repeat your last transmission
SEATCOVERS—Occupants of passenger car, usually attractive female
SET OF DIALS—A CB rig
SEVENTY-THREE—Best wishes
SHACK—Room where radio station is located
SHAKE THE TREES—Act as lead vehicle to decoy any Smokies out of hiding
SHIM—To illegally soup up a transmitter
SHOES—Linear amplifier
SHOUT—Calling someone on CB
SHOUT 'EM DOWN—Look for another station on the band

CB Lingo

SIX WHEELER—Passenger car pulling a trailer
SKIP—Long-distance transmission caused by atmospheric conditions
SKIP LAND—Great distance
SKIP TALKER— A CBer who talks long distances
SKY HOOK—Antenna
SLAUGHTER HOUSE—Channel 11
SLAVE DRIVERS—CBers who take control of a channel
SLIDER—An illegal VFO (variable frequency oscillator)
SMOKEY—The police
SMOKEY IN THE BUSHES—Police concealed
SMOKEY IN THE GRASS—Police waiting at the side of the road
SMOKEY THE BEAR—State police patrol
SMOKEY WITH EARS—Police listening on CB
SMOKE-UM—Same as Smokey
SNOOPERSCOPE—An illegally high antenna
SONNET— A CB'er who advertises products over the air
SOUPED UP—A rig modified to run illegally high power
STACK THEM EIGHTS—Best regards
STAND BY—Wait
STEPPED ON—Signal blotted out by somebody else's transmission
STROLLER—CB'er with a walkie talkie
STRUGGLE—Trying to "break" a channel
SUCKER—A CB rig on the service bench
SUNBEAM—A CB'er who livens the channel with witticisms
SWEEPING LEAVES—Bringing up the rear
SWINGING BEEF—Beef hanging from hooks
TEAR JERKER—A CB'er who always cries the blues
TENNESSEE VALLEY INDIANS—TV Interference (TVI)
THAT'S A COPY—Message received
THIN—A very weak signal
THIN MAN—CB'er with a weak carrier
THIRTY THREE—10-33; This is an emergency
THREAD—Wires in a CB rig
THREES AND EIGHTS—Lots of best regards
THREES ON YOU—Best regards (73)
THROWING A CARRIER—Keying transmitter without talking
TICKER TAPE—FCC rules
TICKS—The minutes that go by (each tick is 60 seconds)
TIJUANA TAXI—Well-marked police car
TIN CAN—CB rig
TOOLED UP—A souped up rig
TRAIN STATION—Traffic court that fines everybody
TREE TOP TALL—Transmission is loud, clear
TURKEY CALL—An intermittent tone generator
TWENTY—Location, short for 10-20
TWIN POTS—A CB'er who has 2 sets from same manufacturer
TWO WHEELER—Motorbike; motorcycle
TWO-WAY RADAR—Radar used from moving police car
UNCLE CHARLIE—FCC
UNGOWA BWANA—O.K.
WALKED ON—Signal blotted out by somebody else's transmission
WALKIN' THE DOG—Checking out what's going on on other channels
WALL TO WALL—Strong signal; or crowded as "Wall-to-wall bears" for well patrolled road
WALL TO WALL BEARS—High concentration of police with strict enforcement traps, etc.
WALLPAPER—QSL cards exchanged by CB'ers with call sign, handle, and home 20 printed on them.
WARDEN—The wife, The FCC
WE DOWN—On the side or listening only; signing off
WE GONE—Same as "We Down"
WHEELS—Automobile; mobile unit
WIERDY—A home-made CB rig
WILCO—Will comply with your request
WIND JAMMER—A long winded CB'er
WRAPPED LOAF—A CB rig in the original carton
WRAPPER—Color; a green unmarked police car would be a "green wrapper"
W.T.—Walkie Talkie
X-RAY MACHINE—Radar
XYL—The wife of a CBer; young lady (from ex-young lady)
YL—Young lady, unmarried CBer
YOUNGVILLE—Young children using the channel
Z's—Sleep

Emergency calls in weather like this are frequent and are quickly answered by police, tow trucks or fellow CBers who relay your message. (Pearce-Simpson)

CITIZENS BAND RADIO DIGEST 115

A blend of the old and the new (RCA)

Fort Worth Texas Highway Department workers posted this sign to slow down drivers entering a highway construction zone.

This gas station owner provides invaluable and quick service to the CB-equipped motorist who has a breakdown on the highway!

Small aircraft use two-way radio for air-to-ground communications. Ranchers use CB for rounding up stray cattle, locating them by air and radioing their location to men on horseback who pick them up.

chapter 10

SERVICE YOU CAN DO YOURSELF

you expect me to mess around in there?

Before you shudder at the thought of what your all-thumbs hands could do to that exotic radio of yours, let's set the ground rules on what this chapter is really supposed to do and not do. These rules are very simple. What it won't do is teach you to become a qualified electronic technician. What it will do is show you how to figure out when you've got a problem and then determine where the problem is. Read on. Even if you're such a klutz with electronics that you need help changing channels on your TV set, there'll be something in these next few pages for you.

(Facing page.) Excellent multimeter kits like this one from Heathkit are satisfying and worthwhile "do-it-yourself" projects.

Probably as many good CB sets come back to dealers for service as sets that actually have a problem. Dismissing those suffering "cockpit trouble"—a switch or other control in the wrong position and the operator unable to figure it out—probably the most common culprit in a case of misdiagnosis is the feedline or antenna. When your radio fails to perform or it's not working up to snuff, you can save yourself a lot of trouble, time and money if you can figure out where your problem lies. Checking out the radio yourself can answer that question for you in most cases.

There are two things you need to check to see if a radio is operating properly; function and performance.

118 Service

The inside of a solid-state two-way radio is a confusing enough place even for an experienced technician. The top and bottom views of this Standard marine radio are typical of what you'd see if you took the cabinet off of your CB set.

This is what a well equipped two-way radio service bench looks like. With the equipment shown, this FCC-licensed technician can solve just about any service problem he's likely to encounter. (Erickson Communications)

Checking function means checking response to controls such as channel selector, squelch or volume. Performance measurements are quantitative and require instruments. If your set doesn't pass the functional checkout and you can't find a cockpit-type screwup, you're probably not going to have to worry about measurement instruments—it's back to the shop!

Wait a minute, you might say—how about opening up the case and taking a look inside for a broken wire or some other obvious problem. You may, if you really want to, but it's not a recommended procedure. In the first place, if the set is still in warranty, opening it up may well void the warranty. Second, the chances that you'll do harm to the set are much greater than the chance you'll find and correct the problem. For example, many sets have speakers or connectors mounted to their cases, and the thin flexible wires connecting them to the chassis or a circuit board are easy to pull loose. Finally, only an FCC-licensed technician is permitted to make adjustments to a CB transmitter, and without detailed service instructions even an experienced service technician can find it tough to tell one part of a circuit from another.

So what tools do we need to find out if a CB set is functioning properly? Number one is a clearly functioning, logical mind. It's the only tool we'll need for some procedures, and it's also the only tool we'll use in all our checks. Here's how to use it for checking out a receiver.

RECEIVER FUNCTIONAL CHECKS

Let's assume we've got a rig in a car, and we don't hear anyone talking on any channel. Is the set getting power as indicated by the lit up pilot light or lamps in the channel selector or meter? Is the transmitter keyed "on" by a jammed mike button or a short in the mike cable? Press the mike "push-to-talk" switch a couple of times to see if the transmit light goes on or the meter reading changes. If this seems to be the problem, unscrew the mike connector (if it's removable) and see if that corrects the problem. If it does you can take the mike apart to see if you can fix it, but beware—most mikes have springs and other little goodies inside that are just waiting to fall out and get lost!

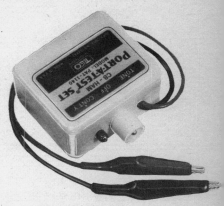

Telco's "Port-A-Test" is primarily a continuity tester—open or short circuits—but is also an "untuned" signal generator for rough checking of amplifier and receiver operation. It can also be used to indicate the presence or absence of voltage.

The next check is to see if you can get any noise out of the receiver by turning the volume (audio gain) control fully clockwise and the squelch control fully counterclockwise. If you get an earsplitting hissing roar, at least part of the receiver is working. If you don't, the receiver may still be OK, but your speaker may be bad. If you're using an external speaker, unplug it. If you're not but the set has an external speaker jack, and you have a speaker or earphone with the right sized jack, plug it in. If you still don't get noise, take the radio in for service. If you are now getting noise, it's time to try for a signal.

Disconnect the coaxial antenna feed line from the back of the radio and stick a short piece of wire into the center hole of the antenna chassis connector. Grip a screwdriver by the metal blade and touch it to the wire. With the set's noise limiter or noise blanker off, you should hear a static-like click in the speaker each time you touch it. Try switching channels. If you still don't hear anything but hiss, there's one last check before going off to the serviceman. Have a friend pull his CB-equipped car alongside yours and see if you can hear him when he transmits. If you can't or if you can but he's weak, it's still off to the shop. At that range a nearby rig will be very strong—even with no antenna at all.

TRANSMITTER FUNCTIONAL TESTS

WARNING! Never key your transmitter without an antenna or dummy antenna connected to the antenna terminal. If you do, you can permanently damage or destroy the output transistor. If you're going to be working on the rig with the antenna disconnected, unscrew the mike connector (if you can) to avoid keying it accidentally.

Before functionally checking the transmitter, we will

assume the receiver is working and seems to be receiving normally. If so, the first transmitter check is the "push-to-talk"—does the transmitter circuit switch *on* when the "push-to-talk" is pressed? If not, the mike or mike cable is the probable problem source. If so, this can usually be confirmed by removing the mike plug and sticking a small screwdriver into its receptacle on the chassis. Shorting the "push-to-talk" pin in the receptacle to ground should turn the transmitter on.

If the transmitter does switch on when the "push to talk" button is pressed, the next thing to determine is whether it is transmitting. Most but not all radios that have meters use the meter to indicate relative RF output on transmit. If so, the meter will read upscale on "transmit" if the rig's putting out a signal. However, if it's one of those meter circuits that reads battery voltage on transmit, there's another way to check for RF, assuming your antenna is bare metal and not plastic sheathed. Wait until dark, and then touch a metal screwdriver blade to the top of the antenna while the mike button is depressed. If there's at least a watt or two of RF, you'll see a spark.

The next thing to check for is modulation. Here you'll need either another CB set or, in a pinch, a portable transistor broadcast band receiver will do. If you've got the CB set fine—just put both sets on the same channel and see if the second set hears good quality audio of normal volume when you modulate the set you're testing. The reason for *not* using another CB set for the previous tests, by the way, is that a nearby reciever will hear a strong signal from a transmitter even when that transmitter is putting out a very weak signal.

If you don't have a second CB set but do have a portable BC set, turn it on and key the transmitter *off* and *on* while you tune the BC set across the BC band. You should find at least one spot where you hear your CB set carrier going on and off (note that it's the actual carrier, not just electrical keying clicks in the receiver). Tune the carrier in and speak into the mike. The signal you hear should have good quality audio, about the same loudness as the broadcast stations you hear on the broadcast receiver. If you have a carrier but little, no or distorted audio, the problem could be the mike or the set. Either way, something needs professional help.

SERVICE WITH INSTRUMENTS

"MUST" INSTRUMENTS

There are two instruments and one device that are so

"There are two instruments and one device that are so useful and so reasonably priced that no CBer should be without them. These are the multimeter, the antenna standing wave ratio meter and the dummy load."

Inexpensive multimeters can often be purchased from hardware or discount stores as well as electronics shops. (Midland)

useful and so reasonably priced that no CBer should be without them. These are the multimeter, the antenna standing wave ratio meter and the dummy load. The multimeter is really a basic instrument for every household, office or motorist. Its ability to measure circuit voltage or continuity can pay for itself many times over.

MULTIMETERS

Multimeters or VOMs (Volt-Ohm-Milliameters, as they are sometimes called) come in a wide range of sizes, shapes—and prices. A very compact model with sufficient accuracy for most uses can be purchased for under $10, while more flexible and accurate instruments range to $100 or even more.

For CB servicing, a multimeter can tell you if the voltage you're supplying to the set is within proper limits and if the current that it draws on both receive and transmit is normal. You can check the mike and antenna cables for open or short circuits. You can also check speakers for open circuits *and* for proper operation. If the speaker is OK, you'll hear it "click" when you check it with the ohm meter.

ANTENNA STANDING WAVE METER

The antenna standing wave meter is primarily an antenna adjustment and checking tool. However, it also can be used as a transmitter relative power meter and a field strength meter that indicates antenna radiation.

The "forward power" setting is the one used when an antenna meter is used to measure transmitter output. In order to be accurate, however, the meter must be connected to a 50-ohm resistive load on the antenna side. Though a perfectly tuned antenna will provide the needed load, antennas are rarely perfectly tuned and are not always conveniently at hand. This is one application where you'll want a dummy load.

Digital multimeters like this B&K Model 280 are much more precise and easier to use than conventional VOMs and still retail for under $100.

Antenna standing wave ratio meters come in a variety of shapes and sizes. Antenna Specialists' M-272 measures absolute power output as well as reflected power and modulation percentage.

Antenna meters like this Heathkit HM-102 are designed to be left in the antenna circuit continuously so have their indicator portion separate from the detector.

Some antenna meters also include an audio detector circuit so you can listen to your transmitter's output with headphones. (Telco)

This Midland antenna standing wave ratio meter can also double as an antenna field strength meter using the telescoping antenna mounted on its rear panel.

An antenna standing wave ratio meter can also be used as a field strength meter by connecting an antenna to one of its terminals. Some compact, low cost (under $20) antenna meters are even supplied with a short telescoping antenna to be used for this purpose.

DUMMY LOAD

A dummy load is nothing more than a 50-ohm resistor (large enough to dissipate its rated power without burning up), mounted on a coaxial connector. Commercial dummy loads for CB use are available for a few dollars, or one can be made by soldering two 100-ohm, 2-watt carbon resistors connected in parallel into a PL-259 coaxial connector. A dummy load is a "must" when you're doing much transmitter testing—it's most inconsiderate to clutter up the crowded CB channels with an empty carrier.

OTHER SERVICE INSTRUMENTS

POWER SUPPLIES

Probably the most useful other "instrument" a CBer could have isn't an instrument at all—it's a 12-volt *regulated* power supply. In a sense it is an instrument since it's a must if you're going to check out mobile rigs anywhere but in the car—a most uncomfortable place to work for any length of time, particularly in cold or hot weather. Of equal importance is the fact that such a supply also makes your mobile rig into a compact, portable base station that can be operated equally easily from home, vacation cottage or motel room.

When selecting a regulated supply, be sure to choose one with sufficient current capability. Two amperes is quite adequate for most CB transceivers, but to be on the safe side you'd better check the manufacturer's specs for "current drain—transmit." Some supplies are also designed to deliver maximum rated current for a limited period of time—if you're a real "ratchet jaw" you could conceivably overheat and damage the supply. A supply with a screwdriver adjustment for voltage is usually a better choice, since output voltage will change as components age and your radio will operate best at its rated voltage. Normal voltage is 13.5 rather than 12 volts, since that's what a car's electrical system actually puts out with the engine running. A supply with "current limiting," though more expensive, is a better choice than one protected only by a fuse because it shuts itself off electronically in case of an overload or short circuit.

One last caution—you want a *regulated* supply, not a battery charger or other bargain power source that

Though inexpensive CB dummy loads are available commercially, it's easy and satisfying to build your own. This one is made of two 2-watt 100-ohm carbon resistors wired in parallel with one lead running through tubular insulation to the center pin and the other soldered to the side of a PL-259 connector. It's usable to 220 MHz.

A *good* regulated 12-volt power supply is a must if you ever want to check out your mobile rig on the bench or operate as a base station. (Midland)

changes voltage when the load current changes. Unregulated "12-volt" supplies can put out 20 volts or more when lightly loaded, for example by a CB transceiver on "receive," and that's enough over-voltage to wipe out most of your rig's transistors and integrated circuits in short order!

SIGNAL GENERATORS

A signal generator is the only instrument that can really tell you how well the receiver portion of your CB transceiver actually hears. Since you'll rarely need a signal generator except when you're doing receiver service work and a good generator—even a well used "antique"—is expensive, you're not likely to want to buy one for your own use. It's a good idea to check out your receiver's sensitivity from time to time, because a gradual loss of sensitivity is usually not obvious to the user. One caution when you're using a signal generator. Unscrew your mike connector before hooking the generator up—4 watts of RF will do the generator no good.

If you are going to get a signal generator, get a good one like this B&K Model 2040. An unstable, poorly shielded generator is of little practical use.

FREQUENCY COUNTERS

Another instrument that's used only infrequently but can be very valuable when it is used is the frequency counter. The counter is not as necessary as it used to be since synthesizers have taken over. Before synthesizers each channel had its own crystal which had to be set on frequency individually and then would proceed to drift—independently of its neighbors—so each had to be rechecked periodically. A synthesized radio uses fewer crystals, hopefully of better quality, and thus is less likely to wander off frequency than its older style counterpart. Nonetheless, a synthesized radio should also be checked for frequency accuracy and stability from time to time.

*One of the more ingenious CB service ideas to come along is B&K's 1040 **Servicemaster**. With the addition of only a signal generator, it checks out all of a CB set's operating parameters.*

RADIO FREQUENCY WATT METERS

A terminating RF power meter is one other instrument that sometimes comes in handy. We already worked up our own version earlier using an antenna meter, but to be sure your 4 watts of 27 MHz RF is really 4 watts, a well-calibrated commercial instrument is a must.

That pretty well sums up those test instruments you will be likely to use or want to use in your CB operations. A fair amount of common sense plus a few dollars invested in some basic instruments can give you a better performing station while helping you learn what makes it perform.

A frequency counter is a must for checking transmitter frequency accuracy and stability. This BOHSEI is usable to 30 MHz.

This six digit frequency counter from B&K signals the user if a frequency beyond its 100 MHz capability is applied.

chapter 11

CITIZENS BAND RADIO AND THE SPORTSMAN

a useful tool that can save lives, too

We've already seen what a valuable tool CB is to both traveler and fixed station operator, but there's another area where CB can provide not only pleasure but also a literally lifesaving link for its users. Outdoor living enthusiasts—campers, hunters, hikers, fishermen—all need some form of communications for those times when things go wrong or even just to help them go right.

The outdoorsman who takes CB radio along is finding more and more friends doing the same thing, and it's a smart visitor to the wild who keeps CB available "just in case." However, out-of-doors CBing is not at all reserved for dire emergencies for, as many are discovering, CBers are using their ability to communicate to enhance their outdoor activities.

Campers, for example, are finding more and more campgrounds equipped with CB radio and advertising that fact with roadside signs. The would-be camper can call in to the campground without even pulling off the freeway to inquire about accommodations or traveling instructions. Once you're in the campground, it's still a useful tool for keeping tabs on the family or simply making new friends.

(Facing page.) This Midland transceiver is a full capability 23-channel mobile and yet is also a portable with its accessory carrying case/battery pack and padded shoulder strap. A removable fold down mount permits use of optional rubber covered whip antenna for short range portable communications.

126 CB For The Outdoorsman

Hunters can make good use of CB by using hand-held portables to organize drives or to tip other members of their party or a nearby group that game is headed their way. Fishermen are using CB in much the same way. Coho fishermen on the Great Lakes in particular are using 27 MHz to let each other know where the schools are running.

RIGS FOR THE OUT-OF-DOORS

One of the basic questions that has to be answered regarding the use of CB in the field is in the choice of a radio. For off-the-road vehicles, campers and the like, that's not much of a problem. The criteria is pretty much the same as in any mobile installation. However, if you're going to be taking your radio out of your vehicle, the crucial factor becomes *portability*.

With rare exceptions a portable CB rig is going to be one of the various brands and models of hand-held transceivers. These come in a bewildering variety of sizes, shapes and capabilities, and making a proper selection from such a field requires your accurate determination of what you really want (and need).

The two conflicting factors that you're going to have to weigh in making your decision are performance *vs* convenience. In an emergency you're going to wish you had a high-powered, many-channeled rig. But if your thing is mountain climbing and you had such a radio, its greater weight and size would have probably forced you to leave it behind at your base camp where it would be useless if you got in trouble on the mountain.

Transmitter power determines range, and the more power there is, the bigger and heavier the batteries are. CB portable transceivers come in three power categories, 100-milliwatt, 1-watt and 5-watt. There's a legal quirk with

Motorcyclists and snowmobilers will find this crash helmet with built-in 100-milliwatt transceiver handy for communicating with fellow enthusiasts while on the move. (Audio Link)

CB-equipped recreational vehicles can call ahead to campgrounds without leaving the highway.

Much greater range is possible from the 5-watt CB hand-held (left), but one of these two 100-milliwatt units (middle and right) would take up less room and be much less weight added to a hiker's or hunter's kit.

the 100-milliwatt 27 MHz transceivers; they do not require a license to operate, and since they are unlicensed, a licensed station is not legally permitted to communicate with one. At the low end of the 100-milliwatt scale are some very cheap radios made as toys and good for little more than toys. Avoid them like the plague. They're essentially useless for what you want to do, and for not too many dollars more you can buy a worthwhile radio. There are some very good 100-milliwatt radios, and if your communication needs are going to be primarily very short range—a few blocks or less (up to a mile line of sight)—they're what you should be looking at. Most are lightweight and compact, with batteries small enough that you won't mind carrying a spare set of them in your pocket or pack. The 100-milliwatt transceivers can be very small, but size limits the number of features. However, one of the most compact models includes a built-in signal tone, a low battery indicator and a two-channel capability—all you will usually need for general short range use. One final note on 100-milliwatt "CB" sets. The FCC has opened a new band at 49 MHz for low power communication devices and eventually 100-milliwatt unlicensed transceivers will be outlawed on 11 meters. Manufacturers have been slow bringing out rigs designed for this new band, but when they do become available, a pair of the new radios will work markedly better than their 27 MHz brothers not only for technical reasons but also because they won't be competing against millions of high power 27 MHz base and mobile stations.

The next step up the portable ladder is to 1-watt. The 1-watt CB transceivers are *really* CB transceivers—they must be the type accepted by the FCC and operated only with a CB license. They're almost always multi-channel and will have features like connections for an external antenna, microphone and power supply. They're larger and heavier

100-milliwatt transceivers are primarily very short range units, transmitting up to a mile line of sight but much shorter distances in heavy brush.

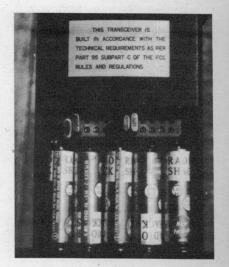
Batteries for many lightweight hand-helds are small enough that you won't mind carrying a spare set of them in your pocket or pack.

Midland's 2-watt, 3-channel hand-held has top control panel, high/low power switch, optional 12-volt power sources for auto or base station use, and holster carrying case with strap.

You must have a CB license to use this 1-watt multichannel transceiver.

This 2-channel, 1-watt transceiver is compact and low cost, ideal for the outdoorsman who needs only short range communications.

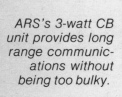

ARS's 3-watt CB unit provides long range communications without being too bulky.

This 3-watt, 8-channel unit by ARS is very compact and can be easily packed into one of the side pockets of a backpack.

than the 100-milliwatt power rigs but are invariably used with a belt or shoulder strap mounted carrying case. They'll do almost everything a mobile rig will do—in fact, some people even use them as mobile rigs because they're so convenient to take along when you leave the car—and the fanciest models even have 23-channel capability. No doubt some will soon be upped to 40 channels, but the same thing we said earlier applies here. The original 23 channels are where the action has been for years and where it'll continue to be. Don't hold out for a 40-channel radio, hand-held or other if your need is now—and don't be afraid to take advantage of a panicky dealer's price cutting on his 23-channel inventory. The money you'll save is yours!

Five-watt (or in-between) portable CB rigs do become unwieldy. A bigger problem (particularly for ratchet jaws) is the short battery life characteristic of the high-power portables. Some do have a low-power switch—use it when you can—but the best advice is to avoid the bigger portables if you move around much, especially if your transportation depends on your own two legs.

One other approach has been used by a few sportsmen who needed more performance than any self-contained radio could provide. They've mounted a compact 5-watt mobile rig, selected for low battery drain and usually modified so the pilot and channel indicator lights were disabled, on a lightweight backpacker's rack along with a couple of high capacity gelled-electrolite storage batteries and suitable mobile antenna. The whole rig is comfortable to carry weighing only about 15 pounds and can be operated in motion. It sounds like a good approach for use at a semipermanent camp setup.

Whatever way you decide to go, give your decision lots of thought, and if possible, discuss it with others who've had experience with CB as an outdoors tool. Probably no other CB application is so demanding in the proper choice of a radio. Choose correctly and you've got a valuable asset for your outdoor living. Choose poorly and you've just wasted your money.

USING CB IN OUTDOOR EMERGENCIES

Because you have so many listeners, CB is the place to be when an emergency strikes. Even in the most remote area of the country the chances are good that someone will be within earshot of a channel 9 "Mayday" when your off-the-road vehicle breaks down or someone in your hiking party falls and breaks a leg.

Channel 9 is always the best channel to go to in an emergency. REACT and ALERT groups cover a great deal of the country, monitoring for emergency calls and keeping the FCC designated emergency channel free of other communications. If your portable has less than 23 channels, one of the channels it should definitely have is 9. If your battery capability will stand it, you should monitor channel 9 for emergency calls from a fellow camper or other outdoorsman in need of assistance or for severe storm warnings and the like.

Since the U.S. Coast Guard refuses to monitor any CB channels because of the amount of background chatter, boaters in areas served by Coast Guard facilities should always have and use marine-band radios. However, if your boating takes you out of range of Coast Guard radio, try channel 13. It's pretty well accepted nationally as the unofficial boaters' channel. Incidently, some CB rigs designed for marine use also have the capability of receiving one or two VHF/FM marine weather channels, a valuable plus when the Weather Service transmissions can be received.

CB INSTALLATION FOR THE OUTDOORSMAN

The outdoorsman has two primary concerns in his use of CB radio, reliability and flexibility. We've pretty well covered these with respect to hand-held portables in the preceding paragraphs, but how about campers, off-the-road vehicles or campsites and their special CB problems.

ANTENNAS

Antennas are a big problem for the CBer who takes his vehicle off the beaten track. Best performance is going to come from a long, semirigid antenna that's mounted high on the vehicle. That's also the antenna that's going to be knocked off by the first low hanging branch. Since maximum performance is important, the answer to the problem is to use a full-size quarter wave antenna mounted as high as possible on the vehicle but to have it on a tilt-over mount which can be quickly and easily folded over for travel. The mount should be located so the antenna can be tilted toward the rear and won't snag on brush or vines on narrow backwoods roads.

Since antennas don't work very well when tilted against the top or side of a vehicle, the well equipped off-the-road traveler will generally install a second CB antenna for backup and travel. This one should be short, preferably base-loaded, flexible and mounted on the vehicle in such a

Midland's 5-watt, 6-channel transceiver can be used with a speaker/microphone, features a jack for an external antenna and top-mounted controls. Note optional flexible "rubber ducky" antenna.

When camping or hiking in the middle of nowhere, it is comforting and reassuring to know that you can communicate with the outside world.

Many CBers who spend a lot of time on backwoods roads prefer the strength of a bumper mount and the effectiveness of a full-sized antenna.

Some CB hand-helds have an accessory antenna jack and can use an external antenna to increase range in areas of poor reception.

way that it's not likely to be damaged in even tight quarters. If you do choose to go the two antenna route, however, don't forget to label your feedlines so you know which antenna is which! You might want to install an antenna selector switch instead of fumbling with connectors when you want to change antennas.

RIG INSTALLATION

The same key factors discussed in Chapter 7 for installing a rig in your car—convenience and usability—also apply to the installation of a rig in your recreational vehicle. Operator convenience is perhaps even more of a factor than it is in a passenger car since off road or back road travel in a heavy vehicle is more demanding of the driver than the "straight and level" travel on a highway. Campers and similar large cab vehicles often lend themselves to ceiling or other mounting locations rather than the dashboard. If you do choose a ceiling mount, remember to install the rig in an area which would prevent the heads of the driver and/or passenger from hitting it during rough travel.

If yours is a pickup truck camper, you'll want to set up a second operating position in the living quarters. You'll need a second 12-volt power cable running to the appropriate spot in the camper, plus some means of bringing the antenna lead-in into the living quarters. This makes a very convenient method of keeping an ear on channel 9 or whatever other channels you may have business on, with the only inconvenience being that of moving the rig back to the cab when you're ready to get on the road again.

RIGS IN CAMP

If you're camping in a tent and you've brought a CB along, your number one concern should be battery life. Of course, if your car is parked close enough to your campsite, you can depend on it for power—just don't forget to

One of the many types of CB mobiles that can be easily installed in a car or RV.

start the engine once a day or so to recharge the battery. As with the pickup truck camper installation, however, you'll want to have the convenience of having the radio in your living quarters if you're going to make much use of it or if you are going to serve your fellow woodsmen as a channel 9 monitor. There's no trick to doing so—an extension power cable, No. 16 or better still a No. 14 wire, plus an extension antenna lead-in will do the trick. As an alternative antenna, you might want to rig up the home brew half wave dipole described in Chapter 5. This is a perfect place for it, and it *will* work better than anything mounted on your car. Either way, unless your car is parked right next to the tent, you should string your car-to-tent wiring 7 or 8 feet off the ground. It's all too easy to miss seeing wires on the ground with a flashlight at night on the way to the outhouse! Another tip is to hang a sign on the steering wheel that says, "STOP! Remember the CB!" If you don't, someone may forget and try to drive to town with the CB still wired to the car.

A mobile unit in camp allows those who stay behind to monitor the activities of those out on the trail.

CB IN YOUR BOAT

The first order of business when planning a marine CB installation is the protection of the rig. Radios dislike water, and the most convenient radio installation ever conceived is useless if the radio gurgles and smokes when it's turned on. Since there are so many possible variations in boat sizes and types, you're going to have to figure out your own details for your situation. Just remember that the water dripping off a swimmer or water skier is just as bad for the radio as wind blown spray. Marine installations call for a specialized marine antenna—a vehicular antenna won't work on your boat unless you have a steel or aluminum cabin top.

Farmers find CB especially useful for coordinating cooperative harvesting activities.

Citizens band radio can be one of the handiest, most useful tools in the outdoorsman's bag. Whether for safety, convenience or simply as a means to meet your fellow sportsman it can't be beat. Next time you head for the outdoors take it along.

1919

FEDERAL
COMMUNICATIONS
COMMISSION

chapter 12

THE FEDERAL COMMUNICATIONS COMMISSION AND YOU

uncle charlie is here to help us!

HISTORY OF THE FCC

The Federal Communications Commission hasn't been around a terribly long time—it was established in 1934—but that makes it one of the older regulatory agencies of the U.S. government. The FCC came into being as a result of the Communications Act of 1934, in which the U.S. Congress brought together the entire communications industry including telephone and telegraph as well as radio. The FCC's purpose is to set up and enforce the regulations under which all telecommunications is to operate since communications is interstate in nature and therefore a federal rather than state or local concern.

The predecessor to the FCC was the Federal Radio Commission, which had been set up by Congress in 1927 to control only radio broadcasting. Prior to 1927 the regulation of radio communications had been under the Commerce Department, except during World War I when the U.S. Navy took it over. U.S. radio regulations of any type don't go back much further than that. Prior to 1912 there was no regulation of radio broadcasting, and everyone did pretty much as he wanted to on the air. Equipment was crude and ranges short, of course, but military, commer-

(Facing page.) Federal Communications Commission Building in Washington, D.C.

This rooftop antenna farm belongs to the monitoring facility at the FCC's main office in downtown Washington, D.C.

FCC Amateur and Citizens Division Chief John Johnston (center) discussing a point with several operators at a hamfest. FCC representatives take this display booth to a CB or amateur radio "get-together" somewhere in the country almost every weekend of the year.

cial, ham and shipboard stations all shared the same frequencies. Sometimes operators of stations in different services helped each other out, but more often they stepped on each other. Thus, as the number of stations grew from the few experimental installations of the early 1900s, it was inevitable that regulation and assignment of licenses and frequencies for operation had to come.

FCC ORGANIZATION

At the top of the FCC's organizational chart are the Commissioners, the seven people who regulate communications for the United States. Commissioners are appointed by the President and confirmed by Congress. Appointments are for a 7-year term, and during that period it is the Commissioners' job to renew and approve or reject all rules proposed by the various bureaus and offices that make up the FCC's working organization.

Regulation of the various areas of communications falls on the various FCC bureaus. These are the:

- Broadcast Bureau
- Cable Television Bureau
- Common Carrier Bureau
- Field Operations Bureau
- Safety and Special Radio Services Bureau

In addition to the bureaus there are also a number of *Offices* in the FCC. The one most important to our activities is the Office of the Chief Engineer.

Citizens Band and the other personal two-way radio services come under the Safety and Special Radio Services Bureau. CB and amateur radio operations are regulated by the Amateur and Citizens Division, while the Aviation and Marine Division has the responsibility for those services. Proposals for changes in the rules regulating a specific service come from the division that is responsible for that service. Such proposals can come about as a result of a petition filed by a citizen or group of citizens or may be developed within a division by its staff.

Before a change in the rules can be adopted, it must be reviewed and approved by all other divisions of the commission that will be affected by it. Then it goes to the Commissioners themselves who may approve it, reject it, or send it back to the originating office or bureau for revision. If its subject is noncontroversial and the Commissioners OK it, that's it—an official announcement is made, and it becomes part of the rules. If, on the other hand, it proposes a change with lots of pros and cons—for example, the proposal to create a new Class E CB band in the amateur 220-225 MHz assignment—it will come out as a "Notice of Proposed Rule Making." An NPRM is a statement of something the Commission is considering doing and invites comments by interested parties to file formal comments up to a certain due date. All the comments received are reviewed by Commission staff members and—eventually—new rules or rule changes are written which take those comments into account.

CB AND THE FCC

Safety and Special Radio Services and Field Operations are the two FCC Bureaus CBers and Amateurs are most likely to come into contact with; Safety and Special Services because it contains the Amateur and Citizens Division and Field Operations because the FCC Field Offices and Monitoring stations come under its jurisdiction. Most of the FCC staff work in Washington—the Amateur and Citizens Division is on the fifth floor of a modern office building at 2025 M Street.

Roughly two dozen men and women make up the staff of the Amateur and Citizens Division—not very many, considering they are responsible for the administration of nearly 300,000 licensed amateurs and many millions of CB licenses! Actual license issuance is done from the Commission's Gettysburg, Pennsylvania, facililty, of course, but the Washington office—(202) 632-7175—still has charge of follow-up on delayed or lost license applications.

The Field Operations Bureau administers the FCC's field offices, 21 in major cities of the continental United States plus Hawaii, Puerto Rico and Alaska. (A list of the current Field Offices and their telephone numbers is included in the back of the book.) Field Offices investigate illegal operations, and interference problems as well as give operator examinations for commercial and amateur licenses.

The Enforcement Division of the Field Operations Bureau has the prime responsibility for seeing that we obey the rules that apply to our particular type of license—Citizens, Amateur, Marine or whatever. To accomplish this the Commission operates 15 highly sophisticated monitoring stations distributed throughout the United States, Alaska, Hawaii and Puerto Rico. These stations operate 'round the clock, monitoring transmissions from the operation of stations throughout the radio frequency spectrum. They issue citations to identifiable stations they hear

FCC's Grand Island, Nebraska monitoring station.

Part of the forest of poles supporting the antennas at Grand Island.

Operating positions at the Grand Island monitoring station. Note the three receivers, tape recorder, frequency measuring equipment and other instruments available to each operator.

Two FCC mobile units, used to locate illegal transmitters.

Another FCC mobile monitoring unit, this one used to compile spectrum occupancy data.

(Facing page.) FCC Official Notice of Violation, something to be avoided but never ignored! If you should ever receive one, be sure to answer it—you're more likely to lose your license for ignoring a violation notice than you are for committing most violations!

operating in violation of the rules, and, in the case of stations not identifiable and/or ones operating illegally, perform direction finding operations. The FCC's DFing capability is quite good, by the way. In the case of a serious violation it's not unusual for three or more monitoring stations to have the violator's location triangulated within a very few miles and within minutes of the time the first monitoring station picks him up and asks for an assist. Then, of course, further action depends on one of the FCC's eight man mobile field monitor units, which uses equipment that can not only trace the signal to the building it's coming from, but can also determine which of several antennas on the suspect's roof is being used!

YOU AND THE RULES

It's no secret that the CB explosion has created a severe enforcement problem for the FCC. With many millions of licensed (and unlicensed) CB stations on the air, the FCC has neither the manpower nor the facilities to properly police 27 MHz.

This doesn't mean they don't try, however, and considering the overwhelming magnitude of the problem, they do a lot more than one might expect. If that sounds like a strong hint that you'd be wise to know the rules and follow them, you're right—it is! As you might expect, the more serious infractions get the most attention and result in the most serious penalties. Any CBer who runs 5 watts on the assigned channels and uses common sense in his choice of language has little to fear from Uncle Charlie. Run a kilowatt with a slider and swap filthy jokes with a good buddy halfway across the country, and the knock on your door may come a lot sooner than you think possible!

One of the worst things a CBer (or any other FCC licensee) can do is ignore a letter or citation from the commission. That in itself—"Failure to respond to an official communication"— is grounds for a healthy fine and/or revocation of your license. It doesn't matter what the subject is—whether it's advising you of a neighbor's complaint about television interference or alleging that you committed a minor rules infraction that you know you are innocent of—you are expected to respond. If you do respond and simply explain your position in the situation, whatever it is, that will often be sufficient to end the matter. If you ignore the FCC's communication even if you can justify not answering in your own mind on the grounds that you were cited in error and it's someone else they are after, you're going to be in trouble.

CITIZENS BAND RADIO DIGEST 137

UNITED STATES OF AMERICA
FEDERAL COMMUNICATIONS COMMISSION

FCC Form 793
December 1973

KINGSVILLE MONITORING STATION

P.O. BOX 632, KINGSVILLE, TEXAS 78363

Form Approved
Budget Bureau No. 52-R0162

OFFICIAL NOTICE OF VIOLATION

WRITTEN REPLY REQUIRED
IN DUPLICATE.

1. Name and Address of Licensee

John Q. Violator
1074 N. Any Ave.
Fairfax, VA

2. Location of Station or Name of Craft	3. Call Sign
FAIRFAX, VA	KXX-0000
4. Radio Service or Class of Station	5. Emission
Citizens (Class D)	A-3

6. Date(s) of Violation	7. Hour(s) of Violation (EST-GMT)	FREQUENCY		
		8a. Authorized	8b. Measured	8c. High/Low (Hertz)
Mar. 27, 1976	5:14 AM CST		27024.97	

WARNING: Certain rule violations, if repeated or willful, as well as failure to reply to this Notice, may result in the imposition of monetary forfeitures. (See Section 510 of the Communications Act of 1934, as amended.) Any of the rule violations, if repeated or willful, may result in the revocation of the station license or suspension of operator license. (See Sections 312 and 303(m) of the Communications Act of 1934, as amended.)

See reverse side for complete instructions.

9. VIOLATION(S)   NON-COMPLIANCE WITH FCC RULES

Section 95.83(a)(7): Transmission of music.

5:14 AM CST . . . so hope you enjoyed the old Miller records, Bright Eyes. This is Night Owl, KXX-0000, on the side.

_____
Carole L. Brown
ISSUING OFFICER
mlb/f

_____
ENGINEER IN CHARGE - LOCATION   KI

April 17, 1976 F
DATE MAILED/SERVED

The knowing and willful making of any false statement in reply to this NOTICE is punishable by fine or imprisonment under Title 18, United States Code, Section 1001.

Adcock direction finding antenna used by FCC monitors to pinpoint illegal stations or locate vessels in distress.

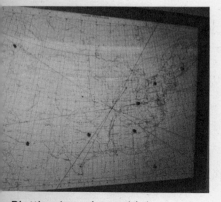

Plotting board on which direction finding information from various FCC monitoring stations is displayed in the initial phase of an effort to locate an illegal transmitter.

(Facing page.) Representative action. Do your best to obey the rules and cooperate with the FCC if you should slip up, and you're very likely to find your name on one of these!

The use of illegal equipment can get you into a lot of trouble. The use of linear amplifiers, though always illegal in the CB service, is still widespread even though the manufacture of CB linears has been banned since January 23, 1975. Any CB operator with an unusually large number of complaints of interference to TV, hi-fi equipment or other radio services is probably going to be suspected of high power operation. Though 5 watts can and does cause some local interference problems, it'll rarely blanket a neighborhood. It's a condition of your license that the Commission can inspect your station when it's in the public interest to do so. Refuse a FCC inspector admittance, and you're looking to lose that license!

VFOs (variable frequency oscillators) or "sliders" are another illegal add-on that can cause you lots of trouble. Some CBers use them just to avoid interference by moving slightly off channel, but from there they slide far enough to land on nearby commercial channels or foul up Class C radio controllers' model planes and boats. To a FCC monitor a slider is obvious and a real red flag—avoid them like the plague!

OUTLAW OPERATORS

The "outlaws"—notice we don't dignify them with the title "CBer"—are those operators who have decided that CB isn't good enough for them and who aren't willing or able to go to the trouble of getting a ham license. No doubt many of the outlaws started out on CB—otherwise they'd be using different frequencies than they do—and some of them surely still hold CB licenses. However, what they are doing now has no relation to any proper CB operation.

Most outlaw operators are using the frequencies just above the legal CB channels, though some are to be found as high as 28 MHz—the bottom edge of the amateur 10 meter band—or, more and more frequently, just inside it. They use call signs such as "HF-1234" or "17W-101" that are actually issued by several large underground groups, and their equipment is typically ham equipment modified to work outside of the ham bands. They run high power—some as high as 5,000 or even *10,000* watts—and talk to other outlaws in other countries as well as across the U.S. Some do eventually graduate to the amateur radio ranks, with groups of them studying together for the amateur exams and even using "their" illegal band for code practice transmissions. The fact remains, however, that such on-the-air activities are about as illegal as it's possible to be

PUBLIC NOTICE

Federal Communications Commission ▪ 1919 M Street, NW. ▪ Washington, D.C. 20554

For recorded listing of releases and texts call 632-0002

For general information call 632-7260

71263

Report No. 1522   SAFETY AND SPECIAL ACTIONS   August 20, 1976 - S

The Commission, by its Safety and Special Radio Services Bureau, took the following action on the dates shown:

August 9 - HOLLYWOOD, FLORIDA, DENNIS F. KOCH, licensee of Citizens radio station KJJ-77?t. Ordered to show cause why licenses should not be revoked for violation of various sections of Part 95 of the rules including Section 95.95(c) of the rules by operating without being identified by its assigned call sign at the beginning and conclusion of each transmission or series of transmissions. (SS-078-76T)

August 10 - SANTA ANA, CALIFORNIA, BRIAN L. TAYLOR, licensee of Citizens radio station KWT-9497. Ordered to show cause why the licenses should not be revoked for violation of various sections of Part 95 of the rules including Section 95.41(d)(2) of the rules by transmitting communications to other radio stations on a frequency reserved for communications between units of the same Citizens radio station. (SS-079-76T)

August 10 - PADUCAH, KENTUCKY, SPENCER H. GRIGGS, SR., licensee of Citizens radio station KEQ-8743. Ordered to show cause why the licenses should not be revoked for violation of various sections of Part 95 of the rules including Section 95.95(c) of the rules for operating without being identified by its assigned call sign at the beginning and conclusion of each transmission or series of transmissions and for violation of Section 97.37(c)(2) of the rules for operating by means of an antenna, with its supporting structure, which exceeded by more than 20 feet the height of the man-made structure to which it was mounted. (SS-081-76T)

with the one exception of deliberately jamming public safety communications.

Don't let yourself get drawn into this type of operation. It's actually a violation of the criminal code of the United States, and the FCC in cooperation with the U.S. Marshalls has been coming down very hard on these people. One recent raid on a group of outlaws in northern New Jersey resulted in their losing over $10,000 worth of equipment, and the individuals involved still face fines and possibly even prison sentences before they're done! If you want more radio than CB provides, become a ham. Your local store, a nearby ham operator or the American Radio Relay League can help you find a class to study with.

CB OPERATING OVERSEAS

CB's legality varies widely from country to country. Canada, for example—"overseas," for the sake of this discussion—has its own "General Radio Service" set up very much like our own CB Service. Other nations flatly prohibit the use of 27 MHz equipment by anyone except government or licensed commercial stations, and levy severe penalties for even possessing a CB rig.

With regard to our good neighbor to the north, our respective governments have worked out reciprocal operating agreements to permit our citizens to operate their CB rigs in each other's country. It's *not* automatic, however—you do need permission. In the case of an American CBer planning to visit Canada, write well in advance of your planned visit to:

Department of Communications
Ottowa,
Ontario K1A 0C8 Canada
Attn: Regulatory Service Branch

Request instructions on how to apply for permission to operate your citizens radio station on General Radio Service frequencies. Since requirements do change, you might try enclosing a photocopy of your FCC/CB license. It's possible you might receive the needed permit without further effort on your part.

Though Canadian customs does have the right to require removal or sealing of your CB equipment if you arrive at the border without permission to operate in Canada, they rarely exercise it. However, don't consider this to be an OK to operate—you could find yourself in hot water if you do. Instead, ask the Customs officer the location of the

One of several racks of sophisticated instrumentation in the spectrum occupancy study van.

nearest DOC Field Office. There you can request and receive operating permission on the spot.

Mexico is an entirely different situation. Until fairly recently, Mexican authorities did grant U.S. CB operators temporary permits to operate their rigs in Mexico. However, so many American visitors abused this privilege—most by not bothering to ask for permission and a few by blatantly illegal activities—that Mexico now flatly forbids any vehicle containing a CB radio from entering the country! If you're driving to Mexico, leave your radio behind or drop it off with a friend before you reach the border. You could, of course, smuggle it over the border in your trunk, but the Mexican authorities would certainly take a very dim view of that, and Mexican jails are not highly regarded as vacation retreats!

If your travels take you to other parts of the globe and you'd like to take your CB set along, start doing your preparatory work well in advance of your departure. If you live near a major U.S. city, you may find an embassy or legation of the country or countries you plan to visit. Call them and ask whether they have a citizens radio service and whether it's possible to get a temporary license to operate there. Chances are that they won't know what you're talking about, but if you press hard enough, they'll either find out for you or tell you where to write to get an answer. A letter to the embassy of the country in Washington is another way to go. Your travel agent should be able to supply addresses. Whatever you do, don't take a radio transmitter into another country without the proper permission. At best it'll probably be confiscated at the border, and the worst could be very bad indeed!

chapter 13

CB CLUBS AND OTHER ORGANIZATIONS

"eyeballing" with your fellow CBers

CB radio is a social activity even if it's not supposed to be according to the FCC. Its current overwhelming popularity is certainly due in large part to the need of people to have someone to relate to even if only through a microphone and loud speaker. The friendly conversations on channel 19 between "Smokey" reports are a prime example of this, though perhaps the 'round the clock chit-chat between base station operators who aren't, after all, trapped in their automobiles is even better.

It's only natural that on-the-air comradery gets carried over into various face-to-face activities. Many of these are informal, spur of the moment—"Hey, guy, I'm going into the next oasis for coffee. Want to join me for a cup?" or "If that's you in the green Kenworth, I'm right behind you in the brown Pinto!"—exchanges, but on the local level CBers who've become friendly over the air have gone much further by organizing various kinds of social or service CB clubs.

Social groups range from loosely organized groups who tend to hang out on a given channel and whose social activities are spur of the moment meetings at a convenient coffee shop or tavern, to highly structured formal clubs

(Facing page.) REACT, Radio Emergency Associated Citizens Teams, is the oldest and largest of the CB service groups and boasts well equipped and well trained public assistance groups in practically every part of the United States.

144 Clubs and Organizations

CB jamborees and hamfests are both social events and excellent opportunities to swap for or buy new gear or accessories.

REACT-equipped and manned stations are often located in quarters provided by local governments with whom they cooperate closely.

Around-the-clock channel 9 monitors include housewives.

whose scheduled activities involve the members' entire families. In both cases a primary or secondary reason for the group's existence may be involved—for example, all members operate SSB or they're all photography nuts who met through CB. Whatever their reason for organizing an off-the-air activity, the clubs do add a significant dimension to their member's lives.

NATIONAL CB PUBLIC SERVICE ORGANIZATIONS

On the more serious side, CB radio has also developed some significant public service oriented organizations. Though there are many such groups in operation at the local level, often in conjunction with or as a part of a basically social CB organization, the best known and most important of these are national in scope.

REACT

Oldest, largest and most influential of all the nationwide CB service groups is REACT—Radio Emergency Associated Citizens Teams. REACT was founded in 1962 as an independent, non-profit corporation. In 1976 REACT rolls carried over 1,500 teams in all 50 states, Puerto Rico, seven Canadian provinces and West Germany. The basic service provided by REACT's more than 70,000 active team members is monitoring channel 9 and providing emergency assistance and road information to anyone who needs it. Highly trained and well equipped REACT volunteers also help out in times of national disasters, major fires or similar events when local authorities need assistance.

REACT is highly organized and operates only on a

team (five or more members) basis. It works closely with civil defense as well as local, state and federal law enforcement organizations. All REACT members pledge that they'll operate in accordance with FCC rules and regulations. Its official statement of purpose is:

1. To develop the use of the Citizens Radio Service as an additional source of communications for emergencies, disasters, and as an emergency aid to individuals;
2. To establish 24-hour volunteer monitoring of emergency calls, particularly over officially designated emergency channels, from Citizens Radio Service licensees, and reporting such calls to appropriate emergency authorities;
3. To promote highway safety by developing programs for providing information and communications assistance to motorists;
4. To coordinate efforts with and provide communications help to other groups, e.g., Red Cross, Civil Defense and local public authorities, in emergencies and disasters;
5. To develop and administer public information projects demonstrating and publicizing the potential benefits and the proper use of Citizens Radio Service to individuals, organizations, industry and government; and
6. To further the above purposes by chartering local Radio Emergency Associated Citizens Teams who will carry out programs implementing the purposes of this corporation on a local basis.

ALERT

Although ALERT (Affiliated League of Emergency Radio Teams) is smaller than REACT, its goals are similar and quite effective in those communities it serves. In addition to providing motorist aid, emergency communications and community service, the Washington, D.C., headquartered organization also does some lobbying in behalf of its membership and the CBing population as a whole.

ALERT claims over 14,000 members and 600 teams nationally. It solicits individual as well as group memberships.

REST-MARINE

REST-Marine does for the CB-equipped boating public what the other two organizations do for the motorist. It was founded in 1971 but has not grown at the same rate as its land oriented cousins. That's unfortunate, as many of the

Channel 9 monitors often supply the first reports of accidents like this tractor-trailer mishap on the Ohio Turnpike.

smaller pleasure boats are equipped only with CB and thus cannot communicate directly with the Coast Guard, which monitors only the marine radio channels. REST-Marine (the REST stands for Radio Emergency Service Teams) has proven effective in the Washington, D.C., and southern Lake Erie and more recently West Coast areas and should continue to expand with CB radio and boating both enjoying unprecedented growth.

NATIONAL CB PUBLIC ASSISTANCE ORGANIZATIONS

ALERT (Affiliated League of Emergency Radio Teams)
818 National Press Building
Washington, D.C. 20004

REACT (Radio Emergency Associated Citizens Teams)
11 East Wacker Drive
Chicago, Illinois 60601

REST-MARINE (Radio Emergency Service Teams)
1039 26th Street South
Arlington, Virginia 22202

OTHER SERVICE GROUPS

Uncounted numbers of local or regional CB public service groups exist. Some are quite good, working quite closely in semi-official capacities with local authorities. Others have, unfortunately, given CB in general a very bad name by adopting police-type uniforms and car markings and infuriating authorities and citizens alike by posing as the law. In general, however, the contributions of the local groups have been as positive as those of the national organizations. If you're interested in finding local activities of this type, your best bet is asking local CBers on the air or inquiring at local dealers.

CB SOCIALIZING

An individual's participation on the CB social scene can be as great or as little as he wishes it to be. "Reading the mail" on local chatter will usually give you insights into what kinds of CB clubs and other off-the-air CB activities are going on in your area and should also point the way as to which specific individuals can fill you in on the details.

News items and ads in local newspapers are another good source for information as to what's happening among your CB neighbors. The larger, better organized groups usually take advantage of opportunities to attract new blood through local news media, and posters in radio shops, supermarkets and other public spots where CBers

---

"An individual's participation on the CB social scene can be as great or as little as he wishes it to be. "Reading the mail" on local chatter will usually give you insights into what kinds of CB clubs and other off-the-air CB activities are going on in your area."

are likely to see them are generally used to herald forthcoming jamborees. Though CB socializing is necessarily a local activity and involves primarily locally based groups, there have been a number of attempts to set up nationwide general interest CB groups as opposed to the emergency service oriented groups discussed earlier. Most of these have fallen by the wayside, but some of the better organized seem to be thriving. At press time a necessarily incomplete list of such groups includes:

American CB Radio Association
Box 1702
Columbia, Missouri 65201

American Federation of CBers
Box 5184
Detroit, Michigan 48235

Brotherhood of CBers
Box 461
New Haven, Connecticut

CB Club of America
Suite 1607
52 Vanderbilt Avenue
New York, New York 10017

Citizens Band Society of America, Ltd.
Warwick, Rhode Island

Single Sideband Club
All America
Box 647
South Orange, New Jersey 07079

Ten-Four International, Inc.
427 Birchwood Avenue
Deerfield, Illinois 60015

The Cross-Country CBers Club
Box 54078
Los Angeles, California 90054

United States CB Club
Washington Crossing, Pennsylvania 18977

United States Citizens Radio Council
3600 Noble Street
Anniston, Alabama 36201

Whether for major public events or during a disaster, CB and amateur radio groups often join forces to provide communications.

FRINGE ORGANIZATIONS

One of the major problems that CB radio has spawned is a significant group of operators who have moved from CB into unlicensed operation on unauthorized channels. Most of this illegal operation is on the 27 MHz band but outside the authorized CB channels on frequencies assigned to other services.

The desires of the operators involved in this type of operation are quite understandable—they simply want to talk long distances without interference, something that's not possible on the crowded CB channels. To do this they are running high power (one outlaw with a 10,000-watt amplifier was nabbed by FCC and Justice Department investigators!) and using converted amateur radio transceivers.

Several organizations boosting this illegal operation have come into existence, and some even issue their members "call letters" to use on the air! This is a good type of activity to avoid, along with the organizations that promote it. The CBer who wants to expand his horizons in two-way radio has the ideal means through amateur radio. A growing number of those who chose the "easy" route have found crime doesn't pay by losing their equipment, receiving fines and even jail sentences!

chapter 14

AMATEUR RADIO

the original citizens radio service is alive, well and growing

Amateur radio is truly the original citizens radio service, and the pioneers of radio such as Hertz, Marconi and De Forest should all certainly be considered "hams." From the earliest days of pre-World War I radio communications when hams with Ford Model T spark coils shared frequencies with equally primitively equipped marine and military stations, amateur radio has grown until today there is hardly a nation in the world that does not issue amateur licenses.

The United States, for example, is approaching 300,000 licensees in its amateur service. The number of licensed hams in the rest of the world is estimated to total about twice that of the United States, and predictions are that the worldwide total will exceed one million hams in the next year or two and may reach two million by the end of the decade.

One of amateur radio's many appeals is its democracy. It has no caste system—bank presidents talk weather with janitors and rap with retired truck drivers about rig problems. King Hussein of Jordan is a ham, as is Senator Barry Goldwater, bandleader Tex Beneke and entertainer Arthur Godfrey. Hams who chase after DX

(Facing page.) Variety is the keynote in many hamshacks. Products of 10 different ham manufacturers plus home brew, World War II surplus and even an obsolete UHF rig formerly used to dispatch taxicabs are at W9JUV's operating position.

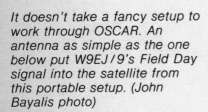

It doesn't take a fancy setup to work through OSCAR. An antenna as simple as the one below put W9EJ/9's Field Day signal into the satellite from this portable setup. (John Bayalis photo)

contacts (foreign countries) will sooner or later have a chat with VR6TC, the only ham on beautiful Pitcairn Island in the South Pacific. He's Tom Christian, direct descendant of Fletcher Christian who led the "Bounty" mutineers to their new life on Pitcairn. Age is no obstacle either—one recent Novice licensee was 5-year-old Neil Rapp of Vincennes, Indiana!

Chewing the rag is the traditional ham activity and probably always will be. Hams are good talkers, though for their "talking" they're just as likely to use fingers as voices. *CW*—chatting via Morse code—is still a popular mode, and some of the serious CW enthusiasts can be heard chatting nightly at 70 or even 80 words per minute! Others prefer radio teletype, TV or a facsimile. DX chasing and experimentation on VHF, UHF or microwave frequencies occupies still others, though you'll find if you go the amateur radio route that most hams end up involved in more than one ham radio interest. VHF/UHF operators, for example, can still work DX using one of the two active ham-built OSCARs (Orbiting Satellite Carrying Amateur Radio). Hams in more than 100 countries have used the ham radio "birds," and several U.S. amateurs have made contact with more than 60 of those countries via the satellites. Another exotic communication mode used by some experimentally minded amateurs is moon-bounce. Richard Hart just contacted Alaska on 144 MHz via the moon, his 50th state on that "line of sight" band from his Delta, Iowa, station K0MQS. By the way, he didn't work all 50 states with the moon's help—some of the others he contacted by bouncing signals off of Aurora displays or trails of ionized gases left by meteors!

AMATEUR RADIO vs CB

Hams, CBers and the public at large all seem to characterize amateur radio and CB as being in opposition to each other. On the surface this would seem to be true—many youngsters who would otherwise have been enthused enough by the romance of radio communication to become hams joined their friends on 11 meters instead. However, there also has to be at least as many people of all ages who started out in CB and—after discovering how much fun there was in two-way radio—buckled down and became hams.

On the other side of the coin, many hams—some of them old timers—are also licensed CBers. CB does, after all, have some privileges that hams don't enjoy, such as its use for business purposes and the fact that a ham's CB rig can be used by any family member without his being present. There are actually good arguments for being both a ham and a CBer, arguments which have been summarized very nicely in the following balance sheet published jointly by the Santa Barbara Amateur Radio Club and the Santa Barbara Citizens Band Radio Club:

Amateur radio is also in space. Shown here (left) is OSCAR 7 (Orbiting Satellite Carrying Amateur Radio) being checked out by aerospace technician Marie Marr and OSCAR 7 project manager W3GEY. Launched on a Thor-Delta rocket in November 1975, (above) OSCAR 7 circles the earth every 114 minutes at an altitude of 900 miles relaying amateur signals between space-minded hams in over 100 countries.

CB vs AMATEUR RADIO*

Each month thousands of Americans are being licensed to carry on radio communications between themselves and others near and far. AMATEUR RADIO and CITIZENS BAND RADIO services offer you a choice of interesting contacts, ways to make new friends and to contribute in meaningful ways through public service and in times of emergency.

AMATEUR RADIO

A name identified with a worldwide fraternity of radio and electronic experimenters and operators licensed at different levels of proficiency and privileged to operate over a broad range of frequencies offering communications opportunities to most any spot on the globe.

Radio Amateurs ("Hams") are permitted to talk to other licensed amateurs most anywhere in the world. Important messages may also be passed to friends and relatives via long-range communications to other countries which permit third-party traffic.

Amateur short-range communications have become highly reliable through the use of hilltop radio repeaters strategically located to relay signals many miles. Amateur Radio enjoys one privilege not granted to all other services—the right to tune for a clear spot anywhere in a designated band rather than to be restricted to one of a set number of fixed channels. Imagine tuning in the South Pole, India, Soviet Central Asia or some remote island on shortwave bands and talking to another ham either in voice or using international code to overcome language barriers and improve reception conditions.

Picture yourself providing assistance in an emergency to a ship at sea or to a stranded motorist on a crowded freeway from within your own car or at home.

Consider the fun of experimenting with a transmitter or some transistorized gadget; building a receiver converter for the band you desire, erecting special antennas, devising control circuits for your operating room. You need not be a graduate in electronics to do this. You learn by following construction articles in radio publications and conversing with other local amateurs at the nearby radio club meeting and on the air.

As a licensed Amateur Radio Operator you can choose a wide range of frequencies from above

CITIZENS BAND RADIO

Citizens Band radio is the most rapidly expanding personal radio service in the world, developed to meet the communications needs of small businessmen, individuals and families across the nation and expanded to fill a wide range of emergency, public service and personal needs of citizens everywhere. Recently expanded to include "hobby" usage, Citizens Band (CB) channels link friends, relatives and strangers by radio for distances permitted up to 150 miles, wherever one travels throughout the U.S. and Canada. Users talk from base stations at home or office or from mobile vehicles on the highways. Mobile radio adds a feeling of safety and provides the convenience of radio-telephone to the CB operator. Convoys of recreational vehicles are able to communicate with each other. Truck drivers have made America aware of the convenience of two-way radio and they make use of the CB channels to alert each other to fuel or rest stops and hazardous road conditions.

The Class D Citizens Band radio consists of 40 communications channels of which Channel 9 is restricted to emergency and motorist-aid use only. In certain areas, Channel 9 is monitored up to 24 hours each day by volunteer organizations. Such service-oriented, emergency, volunteer groups include REACT (Radio Emergency Associated Citizen Teams); VAT (Volunteer Assistance Team); and ALERT (Affiliated League of Emergency Radio Teams). Such groups perform communications services for motorists, municipal and county governments and individuals in need of assistance. CB social clubs exist across the country. Some meet several times each year for "coffee breaks," "pot lucks" (suppers), and other social functions.

SCANNERS AND SHORTWAVE LISTENERS

In the early days of radio, shortwave listeners became fascinated with finding, identifying and listening to radio stations wherever they appeared on the bands. Similarly, today, interest in the

*Courtesy of the Santa Barbara Amateur Radio Club and the Santa Barbara Citizens Band Radio Club

AMATEUR RADIO

the broadcast band to microwaves and use the speed of light to propagate your signal to the area you desire, local or distant—and you can run as much as 1,000 watts of power input to your transmitter, when your signal needs it to reach its desired location.

Amateurs can use many different modes of transmission: Voice—AM, Single Sideband, CW (Continuous Wave—International Morse Code), FM, television, facsimile, teletype and others.

Any adult, boy, girl of *any* age can become an Amateur Radio Operator. Almost 300,000 amateurs are licensed in the U.S. Many are involved in one or more of the following activities:

DX Chasing—Contacting hams in as many different countries as possible and exchanging confirmations; keeping scheduled contacts and making lifelong friends over great distances.

Traffic Handling—Sending and receiving non-commercial messages or providing 'phone patch' contacts between distant parties otherwise unable to talk with one another.

Emergency Communications—Participating in handling messages including health and welfare traffic during actual emergencies or in drills.

Contests—Sharpening one's operating skills by competing for awards for contacting as many stations as possible in a given period; working all states, continents, 100 countries, etc.

Experimenting and Constructing—Building fascinating devices from kits or from original designs, alone or with others.

Satellite Communications—Contacting distant stations by way of OSCAR—Orbiting Satellite Carrying Amateur Radio.

Using Repeaters—To contact other amateurs across the state or around the corner through use of hand-held, mobile or fixed station radios via automatically retransmitted signals.

T Hunts—Locating hidden transmitters with direction-finding equipment in the shortest elapsed time or distance in competition with others or in actual interference cases.

Amateur Radio—Teletype, television, facsimilie, meteor scatter, moonbounce, computers, etc.

Radio Clubs—Hamfests, conventions—getting together with other members of the Amateur fraternity and sharing information, trading ideas and equipment, viewing new products.

CITIZENS BAND RADIO

world about us stimulates individuals to want to tune into the various frequencies allocated to police, sheriff, highway patrol, fire, ambulance, aircraft, Citizens Band and Radio Amateur. These services occupy different frequencies, some within the same band. One may monitor them through the use of a scanning receiver which can be set to "lock on" to any given channel as soon as it is in use, thereby permitting the listener to hear interesting messages—as long as one merely *listens* to them.

Many CB radio owners own scanners to permit them to participate in public service work as a member of a team as outlined above. It is not necessary to be licensed to own a "scanner." However, most scanner owners are also interested enough to become licensed for the type of radio transmitting they wish to enjoy.

The CITIZENS RADIO SERVICE license is issued by the FCC. Applicants must be 18 years of age or older, but any member of the Licensee's family may use the radio. Most of the CB social functions are enjoyed by families as a whole. The Federal Communications Commission licenses Class D operation of transmitters on frequencies in the 27 MHz region ("low band") using the maximum power of 4 watts output AM or 12 watts P.E.P. (Peak Envelope Power) on Single Sideband. The operator's antenna at his base station may be as high as 60 feet, and he may communicate as far as 150 miles.* Short range mobile communications normally consist of 5- to 50-mile distances. Atmospheric conditions often permit signals to be heard thousands of miles away but such contacts are neither dependable nor legal.

Equipment available to the CB operator costs from $100 up. No technical knowledge is required to either purchase, license or operate CB radios. Licensing is a simple registration process costing $4 for a 5-year term. Applications are currently pouring into F.C.C. offices at a rate of over 400,000 per month. Soon, all auto manufacturers will offer dash mounted CB radios as optional equipment on their cars. With hundreds of thousands of CB radios and operators licensed, in all likelihood the antenna you see on the vehicle you pass indicates the presence of Citizens Band radio. Other Citizens Band Service bands are available for business activities of the licensee or for the control of remote objects.

---

*For full rules and regulations regarding antenna height, see the FCC Rules and Regulations in the back of this book.

K5TYP strings an antenna in the wreckage left behind by the tornado in Biloxi, Mississippi. (U.S. AF photo)

Amateur antenna installations can be quite elaborate. This is W9NZM in suburban Chicago.

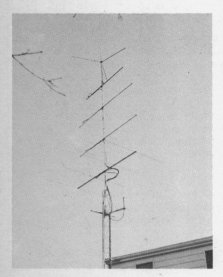

The antennas stacked on K3MWV's tower cover 5 VHF/UHF amateur bands.

HAM RADIO'S HISTORY

Although at the very beginning all the people operating radio stations were "amateurs," most of those primitive stations fairly quickly assumed some sort of official government or commercial status. Those that didn't— those belonging to pioneers of the early 1900s who were playing with radio purely for fun—were the first true ham stations.

With all the early transmitters, "spark" rigs, sending out bursts of static in Morse code, those early stations shared the same long wave portion of the spectrum, and intolerable interference was the result.

When the first amateur licenses were issued in the United States in 1912, amateurs were given the "useless" frequencies above 200 meters (*all* shortwave and higher frequencies, starting just above the present AM broadcast band!). The amateur population then numbered in the hundreds, but by the end of 1916, 5,424 citizens held amateur licenses. In January, 1917, amateurs succeeded in relaying a message cross-country in a single night, quite a feat when you remember that the maximum range of a primitive station of the period was only a few hundred miles. Relays (hence the American Radio Relay League's name) were a must for a radio message to travel any appreciable distance.

Then came the war, and amateur radio was shut down while hundreds of ham volunteers donned uniforms to staff burgeoning military communications needs. When the war ended the Navy wanted to continue its wartime control of radio communications, but after a fight led by the ARRL (American Radio Relay League), amateurs went back on the air in October, 1919. By 1922 over 15,000 amateurs, most still using spark but a few experimenting with vacuum tube transmitters, were on the air and a few were even being heard in Europe. It wasn't until November 27, 1923, however, that the first transatlantic two-way contact by amateur radio actually took place.

That contact not only woke up the amateur world—it woke up the communications industry, too. That historic contact took place on 110 meters, approximately 3 MHz, in that "useless" shortwave spectrum to which the amateurs had been banished years previously. The radio spectrum was hurriedly reorganized, and the following July 24, amateurs were given specific bands at 80, 40, 20 and 5 meters.

Amateur radio continued its growth, doubling from

15,000 in 1922 to 30,000 in 1932 and reaching nearly 300,000 licensed U.S. hams in 1976. At the same time amateur radio has been growing in numbers, it has also been growing technically—amateurs were experimenting with television in the mid 1920s, and many of the advances in equipment and antenna design modern engineers take for granted came from the basement or attic workshops of inquisitive hams.

WHO CAN BE A HAM?

Anyone can become a licensed amateur radio operator. Unlike CB there's no age limit—Neil Rapp of Vincennes, Indiana, recently passed his Novice Class license exam and is talking to the world as WN9VPG. It took him three tries before he passed, but then Neil is only 5 years old! There are many women hams, many of them licensed in grammar or high school, who keep up their on-the-air activities along with housework or careers.

Physical handicaps are no bar to becoming an amateur, and in fact, hams have developed various aids to help blind or deaf hams operate and even maintain their station equipment. For many years one of the Midwest's top DX chasers operated from an iron lung, a badly paralyzed victim of polio. Because amateur radio has such therapeutic value and, conversely, handicapped hams have made so many contributions to the amateur radio service, the FCC has long made special provisions to permit them to take the amateur license exams.

Many would-be hams shy away from amateur radio because they think it's too tough for them. Nonsense! Anyone who's been active in CB for any period of time and has been the least bit curious about his equipment and antennas has probably picked up enough elementary theory to pass the Novice exam! Of course, there's still that

Age is no obstacle in amateur radio. WB3ZEA passed the Extra Class license exam (ham radio's toughest) March 19, 1976, just a year and a week after receiving his Novice license. He's 13 years old!

Unlike CB, there's no age limit for becoming a ham. Neil ("Rusty") Rapp was only 5 years old when he passed his Novice exam and became WN9VPG. (Photo by Paul Willis, Vincennes Sun-Commercial)

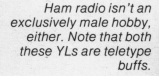

Ham radio isn't an exclusively male hobby, either. Note that both these YLs are teletype buffs.

terrible code... but the five words per minute (that's more than two seconds to recognize and write down each character) that the Novice or Technician Class licenses require can be achieved by most people in a couple of weeks if they practice one hour per night. Look at the code in the accompanying box—is it really so tough?

Morse Code

A	didah •—	N	dahdit —•	1	didahdahdahdah •————	Period:	didahdidahdidah •—•—•—
B	dahdididit —•••	O	dahdahdah ———	2	dididahdahdah ••———	Comma:	dahdahdididahdah ——••——
C	dahdidahdit —•—•	P	didahdahdit •——•	3	dididahdah •••——	Question mark:	dididahdahdidit ••——••
D	dahdidit —••	Q	dahdahdidah ——•—	4	didididah ••••—	Error:	didididididididit ••••••••
E	dit •	R	didahdit •—•	5	dididididit •••••	Double dash:	dahdidididah —•••—
F	dididahdit ••—•	S	dididit •••	6	dahdidididit —••••	Colon:	dahdahdahdididit ———•••
G	dahdahdit ——•	T	dah —	7	dahdahdididit ——•••	Semicolon:	dahdidahdidahdit —•—•—•
H	didididit ••••	U	dididah ••—	8	dahdahdahdidit ———••	Parenthesis:	dahdidahdahdidah —•——•—
I	didit ••	V	dididah •••—	9	dahdahdahdahdit ————•	Fraction Bar:	dahdididahdit —••—•
J	didahdahdah •———	W	didahdah •——	0	dahdahdahdahdah —————	Wait:	didahdididit •—•••
K	dahdidah —•—	X	dahdididah —••—			End of message:	didahdidahdit •—•—•
L	didahdidit •—••	Y	dahdidahdah —•——			Go Ahead:	dahdidah —•—
M	dahdah ——	Z	dahdahdidit ——••			End of contact:	dididididahdidah •••—•—

One caution: NEVER think of the code as "dot-dash, dash-dot-dot-dot" and so on. Morse code is *sounds*, not words, so "A" is "di-dah,"; "B" is "dah-di-di-dit" and on through the alphabet. If you start out thinking how code *sounds,* it'll come fast for you.

Of course, a Novice license is pretty restrictive—at least by ham standards—but a Novice like Neil can use a 250-watt transmitter with VFO and any antenna he wants to while operating on portions of the 3.5, 7, 21 and 28 MHz amateur bands. A Technician (still five words per minute code plus a tougher theory exam) can do everything a Novice can plus operate any mode—CW, voice, radio-

Even hand-held portables like this Genave come equipped for making phone calls.

Relatively few hams have seen an amateur license that looks like this one. It's on a self-mailing form that was introduced in mid 1976, and it saves about a week in processing time.

Ham radio equipment comes in all shapes and sizes. The Kenwood TS-820 is a CW/SSB transceiver that covers all the amateur bands from 1.8 through 28 MHz.

Many hams still build their own equipment though much of what they build today comes from kits. This is Heathkit's Model HW-2036 synthesized 144 MHz transceiver. Note the Touchtone® pad mounted on the microphone, which enables the user to dial up his own telephone number through an appropriately equipped repeater.

teletype, television—with up to 1,000 watts of power on the 50 MHz and higher frequency amateur bands.

General Class is the next step up the ham ladder, and with a General license a ham can operate on portions of every amateur band using all modes permitted on those bands and up to 1,000 watts of power. The two highest amateur license grades, Advanced and Extra, give the holders access to additional parts of several of the most popular ham bands as an incentive for the individual amateur to maintain his proficiency and upgrade himself.

If you're interested in becoming a ham, there are a number of ways to go about it. Local ham clubs and high school and college evening programs all offer courses on becoming a radio amateur. The *American Radio Relay League* has an amateur radio training course as does *Ham Radio Magazine*—their addresses are at the end of the chapter. Individual amateurs are often interested in helping newcomers into ham radio, and a notice on the bulletin board at your local ham radio or CB shop will often put you in touch with other nearby would-be hams you can work with. Try it!

Wait a minute! Isn't ham radio a terribly expensive hobby? What about those fancy stations you see in the ham magazines—don't their owners have thousands of dollars invested in them?

Amateur radio, like photography or restoring a vintage car, is as inexpensive or as costly as *you* choose to make it. You can put together a ham station that could literally reach any corner of the world for under $100! Needless to say, you'd be using older model second- or third-hand equipment and lots of elbow grease and ingenuity, but it could be done. On the other hand, the most expensive current ham transceiver lists for well over $2,000—and you could spend at least as much as that on the antenna

Field Day experience helps amateurs prepare for real disasters. Here WA0CPW surveys the destruction from the Rapid City, South Dakota flood.

Every June hams take to the field to test their abilities to operate under simulated emergency conditions. This California group is setting up its station, W6PIY/6, for the annual "Field Day" competition.

A relatively spartan hamshack can still reach 'round the world. The certificates on W1DTY's wall show that he's confirmed contacts with more than 180 countries on all continents.

A ham's "shack" is where he wants it to be. WA9KRL operates the Heathkit transceiver he built from his car.

system to go with it. For the ham on a limited budget, judicious shopping at hamfest flea markets should get him everything needed for an effective ham station for a few hundred dollars, more or less.

AMATEUR RADIO PUBLICATIONS

Radio amateurs have five monthly magazines, one monthly tabloid (newspaper style) and a weekly newsletter to keep them up-to-date on happenings in the amateur radio field. In addition there are many special interest ham publications for enthusiasts of radio-teletype, television, amateur satellites, DX and similar specialties.

Of the five magazines QST, the journal of the American Radio Relay League, is by far the oldest (1915) and has the largest circulation. Its mix includes first class technical articles, reports on League contests and other activities, and regular columns for specialists in particular amateur areas of interest. Second largest of the amateur radio publications is *Ham Radio*, which is directed entirely toward the technically minded ham and is so highly regarded that many non-ham engineers and scientists subscribe to it. *73 Magazine* can be classified as a general interest publication, and in its editorial pages you're likely to find a travelogue of a "DX-pedition" to some obscure island, a humorous piece on teaching ham radio to your wife, news of a FMers conference and some technical pieces to boot. *CQ* is also a general interest book, with considerable emphasis on contests, awards and operating activities—it's also the skimpiest of the five. *Ham Radio Horizons* is the newest (January, 1977) of the amateur radio magazines and is directed specifically at the newcomer or potential newcomer to amateur radio. In addition to beginning level technical articles and general interest material, it will also have some CB and shortwave listening coverage.

HR Report is a weekly 4 page newsletter that gives its readers by far the fastest access to current FCC, ARRL and industry happenings. Its concise, current content is very popular with the more active hams as well as industry and government officials interested in amateur radio—though its weekly first class mail delivery makes it the most expensive of all the amateur publications. *Worldradio* also does a pretty good job with current news-type items, though hampered by its monthly publication schedule. It also features special interest columns.

All five of the monthly magazines are distributed by some magazine stands or by radio stores that have ham radio clientele, and *Worldradio* will send a prospective subscriber a sample copy on request. If you're serious about becoming a ham, you'll find all the ham publications worth your while. Subscription information appears below.

A ham needn't have a car to operate mobile. A ham portable 2-meter like this Drake TR-33C can reach 50 miles or more with the aid of an amateur built and maintained repeater station.

QST $9.00 a year
The American Radio Relay League
225 Main St.
Newington, Connecticut 06111

The ARRL also publshes a number of excellent books on almost every phase of amateur radio.

Ham Radio Magazine $10.00 a year
Ham Radio Horizons $10.00 a year
HR Report $16.00 a year
Ham Radio
Greenville, New Hampshire 03048

Ham Radio also publishes several worthwhile amateur radio books, and is probably the largest distributor of electronic publications (including CB, SWL, micro computers and the like) in the world.

73 Magazine $10.00 a year
73, Inc.
Peterborough, New Hampshire 03458

73 also publishes books and code tapes for amateur radio.

CQ $ 7.50 a year
Cowan Publishing Corp.
14 Vanderventor Ave.
Port Washington, L.I., New York 11050

Cowan also publishes *S9 Magazine*, the largest circulation magazine for CBers.

Worldradio $ 5.50 a year
Worldradio Associates
2120 28th St.
Sacramento, California 95818

The wind is a traditional power source that hasn't been overlooked by Field Day participants.

Bonus points are given Field Day stations for contacts using unusual power sources. A bicycle driven generator supplies power for W6PIY/6.

chapter 15

OTHER PERSONAL TWO-WAY RADIO MODES

CB and ham radio aren't your only options

Though it's citizens band that's recently been hogging the spotlight and three generations of hams have boosted their service since radio communication began, those are not the only options for the individual with a need or desire to talk by wireless. Pleasure boat and private aircraft owners are also very active communicators, though of course only as an adjunct to their primary boating or sailing activity.

Those are still not the only alternatives open to the individual interested in using the radio spectrum or curious about what goes on in it. There are also some very limited range, no license modes of two-way radio communication permitted by the FCC, and for the individual more curious about how other people are using radio but not too anxious to be an active participant himself, there's always the ancient and respected sport of shortwave listening. For starters, however, let's consider marine and aircraft radio.

MARINE RADIO

Ship-to-shore communications was one of the very first practical applications of the infant radio industry and remains so to this day. It's come a long way, of course—the

(Facing page.) Though the "driver's seat" looks a good deal different, most of the same installation considerations that apply in a 4-wheeler or big semi also affect a radio used in a boat.

Other Radio Modes

That a power yacht like this Trojan F-32 Sport Fisherman (right) would have an impressive array of communications equipment is not surprising, but even a number of day-sailers like these (below) in an Ohio marina are now equipped with some sort of two-way radio.

Motorolas Triton 55/75 synthesized VHF marine radiotelephone is designed for use on larger pleasure boats or even commercial vessels.

You don't have to pilot a big power boat to enjoy the safety and convenience of VHF marine radio. Hand-held transceivers like this Standard provide surprising range from the deck of a sailboat or even a canoe.

Titanics's "CQD" when it hit that deadly iceberg in 1912 was transmitted by the static-like noises generated by a spark-gap transmitter, while modern merchant and military vessels use satellites for both communications and navigation.

The small boat owner has neither the need nor bankroll for such an exotic solution to his communications needs, however. Unless he's one of the adventuresome breed that braves the high seas in his pleasure craft (or is a multi-millionaire whose yacht is a *ship*, not a *boat*), his radio needs are for simple, reliable short-range communications.

These needs are accommodated by the marine VHF/FM channels around 160 MHz, a portion of the spectrum that lies between the VHF police and fire band and TV channel 7. Up until fairly recently private boaters were limited to the 2 MHz marine band, practically the other end of the radio communications spectrum, but that's phased out now and a good thing—propagation at that frequency was such that on a cold winter night it was often easier to contact a Coast Guard station 500 miles away than the one whose lights you could see on the horizon!

VHF/FM is line of sight comunications like TV or broadcast FM which makes it ideal for coastline or inland boat-to-boat or boat-to-shore links. Over water where there're no obstacles such as trees or buildings, even low power transmitters can achieve impressive ranges. There are over 90 channels assigned to the Marine service, though very few small boaters have occasion to use more than a few of them. Channel 16 (156.8 MHz) is the distress and calling channel, and channel 6 (156.3 MHz) is designated as the safety channel. Various other channels are assigned for ship-to-ship, Coast Guard or other specific uses.

For the yachter who does intend spending much time out of sight of land, the longer distance shortwave marine bands are a must. Radios for this service operate on SSB on various designated frequencies in the 4, 6, 8, 12, 17 and 22 MHz bands in order to be able to communicate *with* almost any part of the world *from* almost any part of the world. Marine radio operation in the shortwave spectrum is a science in its own right—for our purposes it's perhaps also sufficient to say that the marine shortwave bands provide interesting hunting grounds for shortwave listeners.

So far we haven't mentioned CB for shipboard use in this chapter. There's a good reason for the omission. CB usage in urban areas, even with the expansion to 40 channels, is so heavy that it cannot be considered a primary communication mode. Interference problems are so severe, in fact, that after some unsatisfactory attempts the Coast Guard flatly refuses to take part in CB communication! This doesn't mean you're doomed if CB is your only means of reporting a distress situation—the Coast Guard will respond to distress calls relayed by CB monitors—but you are out of luck if you'd like to report your problem directly to them.

CB is also widely used by boaters in inland areas or along remote coastal reaches where Coast Guard and other VHF/FM facilities don't reach. It is also invaluable for fishermen and pleasure boaters who want to share their activities with others having common interests.

Like all the other FCC regulated radio services, a license is required for a shipboard radio station. Applications should be available from your marina or any marine equipment dealer who handles marine radio equipment. Forms can also be requested from the nearest FCC Field Office listed in the back of this book. Ask for FCC Form 502. You'll also need an operator's license called a "Restricted Radiotelephone Operator Permit" to operate a shipboard transmitter on the marine bands. That license you get by filling out and mailing in an FCC Form 753.

For the boater on a limited budget, marine FM rigs like Genave's 10-channel Marine Mate 100 offer excellent performance at a nominal price.

A boat owner looking for marine radio equipment should seek the advice of a competent, well-established marine dealer or a two-way radio distributor who specializes in marine radio.

The RT-55M by Emergency Beacon is a 55 channel synthesized radio that can also be used with an external hailing speaker.

If you do plan on using CB on your boat, the lack of a metal groundplane to mount it on will require use of a special type of antenna like this Avanti AVS8.

Pearce-Simpson's Capri 25 not only covers the necessary VHF marine channels but also includes both VHF weather channels and a broadcast band receiver.

This sports fishing boat carries antennas for low frequency SSB, VHF/FM and CB communications plus radar.

Some very compact radios like this 12-channel SR-C8075 from Standard make marine installations possible in very cramped quarters.

An LED channel indicator highlights the Regency Model MT-55 synthesized marine radio which covers all 55 domestic and international marine FM channels.

Hy-Gain's Hy-Seas 55 provides coverage of all 55 marine channels.

If your boat is harbored in an area where break-ins are common, you'd be wise to consider one of Trans-Comm's portable packages. The M-2940 combines a Royce VHF marine transceiver and a Regency Whammo-10 programmable VHF scanning receiver.

Equipment for the marine VHF/FM band looks very similar to CB equipment. It has similar controls and, in many cases, is made by the same manufacturers. You'll find one big difference when you operate it, however. It's clearer sounding at much greater ranges, both because of the lack of interference from other stations and because it uses frequency modulation.

Though there are not as many different makes and models offered for the VHF marine bands as there are for CB, there is ample variety. For the sailboat user who has no shipboard power, a compact hand-held transceiver can be quite satisfactory—particularly if he installs a mast-top antenna that can be plugged into the hand-held for off shore cruising. Slightly more sophisticated low power transceivers operating off internal or external battery packs are also available.

For power boats or larger sailboats with ample battery power, the larger, many channel radios are recommended. They're more expensive, of course, but the improvement in performance is worth the cost.

One last comment about marine radio. It is, after all, a tool used by boaters for safety and other *necessary* communications pertaining to boating operations. Unfortunately, bad operating habits have been increasing at an alarming rate on the marine channels to the point where a combined user/industry advisory group recently reported the following severe abuses of marine VHF/FM:

- Unnecessary and over-use of channel 16 (the emergency channel).
- Unnecessary and/or overly long conversations.
- Use of incorrect or improper channel.
- Interruption of other traffic with non-emergency calls.
- Procedural faults (identification, call signs, etc.).

Since it's your, the boater's, life that may depend on proper use of the marine channels, it's you who should be using them properly.

If you're planning ocean travel, you'll need long range HF/SSB like Genave's GSB-100 to supplement your VHF communications equipment.

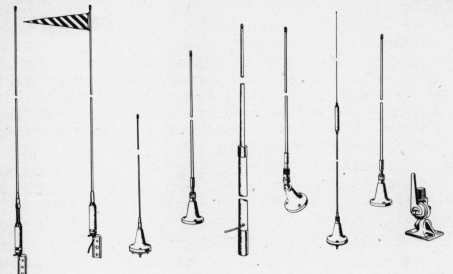

As in the case with CB, marine VHF two-way radio antennas come in a wide variety of shapes and sizes as shown in this group from Antenna Specialists.

AVIATION RADIO

For the individual who flys a light plane, a two-way radio is a must. In the first place, you must have —and use—a radio if you're flying a plane near any major urban area. The Federal Aviation Agency forbids aircraft not equipped with radio from operating in high density areas. Secondly, it's just good sense. A pilot flying cross-country in an airplane at 120 miles per hour or more can encounter all kinds of weather extremes in a few hours—or sometimes minutes!

Aircraft communications channels are located in the 120 MHz portion of the spectrum. Aviation navigation aids such as omnirange (VOR), are located between 108 and 118 MHz, tower and approach/departure control just above 118 MHz, channels used by private aircraft for com-

Whether flying a near antique like this 1947 Cessna 140 (left) or a sparkling new light aircraft like Cessna's 310 II, (right) a private pilot is going to need two-way radio.

Genave's Alpha portable aircraft radio is an ideal answer for older Piper Cubs, Aeroncas and other models without electrical systems or as a backup for the built-in radio system.

When listening to aircraft communications channels, you'll hear little of the banter that's common almost everywhere else in the spectrum.

Shortwave listening can be done with equipment as elemental as a broadcast receiver with a shortwave band to very sophisticated (and expensive) professional type receivers such as Drake's DSR-2.

municating with each other or airport facilities for information or instructions around 123 MHz, and in-flight information, airline company frequencies and the like above that.

The communications mode for VHF aircraft communications is amplitude modulation, just as it is for CB. In addition to the VHF aircraft band there is also a UHF aircraft band at 225-400 MHz, strictly military, and several shortwave aircraft bands used by international airlines. There are also some old range stations in the 200-400 kHz long wave band, but these are used more today for transmitting recorded weather information than for navigation.

As you might expect, communications on the aircraft frequencies are brief, to the point and deadly serious. Unlike the marine channels, where a certain amount of casual conversation seems to be the rule, friendly chit-chat is an absolute "No-No" to people who fly planes.

If you're planning to become a pilot, aircraft radio will be a very important item on your training curriculum. You will be tested on operating procedures and use of aircraft radio as part of your pilot's exam. As is the case with a marine radio station, you'll need a Restricted Radiotelephone Operators Permit before you solo. Until then, it's OK to operate the aircraft since you're doing it under the supervision of the instructor who is already a licensed operator. Use FCC Form 753 (FCC Form 755, if you're an alien) to apply for your operator's license. If you're fortunate enough to own your own plane, you'll also need a station license for it which you must apply for with an FCC Form 404.

SHORTWAVE LISTENING

Though shortwave listening is not the same thing as taking part in your own two-way radio communication, a very similar fascination for the wonders of radio is certainly a part of it. Most hams probably start out as shortwave listeners, though many confirmed SWLs are content to listen to what other people are doing on various parts of the radio spectrum.

Until recent years almost all SWLing was done in the shortwave bands between say 2 and 30 MHz. International broadcast monitoring combines lots of entertainment with the thrill of the chase, for when the novelty of Radio Moscow, the BBC, Deutsche Welle and Radio Netherlands—all producers of booming signals that can be heard well on literally any radio with a shortwave band—

wears off, the challenge of hearing weaker and weaker signals from more and more obscure places takes over.

International broadcast is only a part of the action on shortwave, however. There are the marine and aircraft channels, ham bands, military operations, navigation and time signals, and much much more to challenge the shortwave listener. What he hears is limited only by his equipment—a *good* general coverage shortwave receiver and antenna are a must—and his ability to use it, plus his ingenuity in figuring out what he's hearing!

Another form of shortwave listening has become almost predominant recently—the monitoring of public service activities in the VHF and UHF spectrums. This is partly a result of technical innovation, which has made good quality receiving equipment for these frequencies available at consumer prices, and partly due to increased public awareness of the drama available on police and fire radio channels.

Scanning receivers are the most popular form of monitor. These are multi-channel radios which are set (usually by purchasing a plug-in crystal) to the various police and fire department frequencies in the user's locality. Then, on command the receiver switches itself successively from channel to channel, passing by those that are not active but pausing to listen to any that are. Though most of what's overheard is routine and uninteresting, there's usually enough excitement in a given period of time to make such eavesdropping entertaining.

Though most of this type of monitoring is confined to the so-called public service channels, there are also similar monitors available for the aircraft band. The more sophisticated public service monitors are multi-band and cover 30-50, 150-170 and 450-470 MHz, and some are even synthesized, enabling the user to program any frequency or frequencies he wishes into the unit and avoid having to buy crystals.

CHOOSING EQUIPMENT

As in choosing a CB set, there are a lot of things to consider before laying out your money for a shortwave receiver or scanner. What you really want out of your listening activity is paramount—if you'll be content listening to music or commentary from one of the stronger international broadcasters, you don't need a sophisticated communications receiver. If the only public service activity you plan is eavesdropping on your own city's police and fire, you

This synthesized SWL receiver tunes 500kHz through 30 MHz for AM, SSB or CW and can operate from an internal flashlight battery pack as well as from 115 VAC. (R.L. Drake)

Drake's SPR-4 general coverage shortwave receiver can be digitally tuned with the FS-4 accessory synthesizer.

Eavesdropping on police, fire or other VHF/FM services is the current SWL rage. Regency's ACT-C4 is a 4-channel scanner that covers both VHF high and low bands along with UHF.

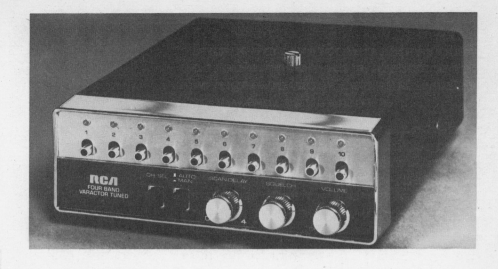

RCA's four band scanner can be set up on as many as 10 preselected channels.

The Bearcat 8 is a three band, 8-channel scanning receiver.

don't even need a scanner—a 2- or 4-channel manually-switched VHF receiver will do, since the police dispatcher will advise the police cars whenever there's a fire call and then you can switch over to the fire channel.

Good used equipment is often available at very modest prices. Do some shopping, and enlist the aid of someone familiar with the field before investing a lot of money. Shop wisely and a small investment can bring literally years of pleasure—good listening!

WARNING! One pitfall threatens the unwary listener to or user of the airwaves—the Communications Act of 1934. Under the terms of the Act, the secrecy of all radio communications are protected by law with only two exceptions, broadcast and amateur radio.

This means that though you can listen to anything that goes on anywhere in the radio spectrum what you hear is considered privileged information. Legally, you cannot repeat what you hear to another person or use it for your profit. Keep this in mind as you listen. Though prosecutions for violations of the Act's security provisions are rare, discretion is certainly advised!

UNLICENSED RADIO COMMUNICATIONS

No, we're not talking about the antics of the outlaw "HF" or "W" groups that operate high power surplus or converted ham gear without benefit of a license on whatever frequencies they choose. Our brand of unlicensed communications is sanctioned by the FCC—Part 15 of the Rules, if you want to see for yourself.

There's quite a variety in what's permitted under Part

15, but it all has one common denominator—low power. For example, the low power, hand-held 27 MHz transceivers we looked at previously when we discussed portable equipment are authorized under Part 15 along with garage door openers, wireless microphones, and some telemetry transmitters.

Portable transceivers operating on 27 MHz can be useful communications tools but, at least in the low powered unlicensed versions, have been more looked on as toys. Many of them are indeed toys, with performance so poor that they don't even make good playthings. Among the better low power models the user still has problems. First and foremost, since they operate on CB channels, their range is limited by other much higher power CB activity, and there's no shortage of that. Secondly, you can't legally talk to a licensed CB station with one even if he can hear you.

Eventually these limitations are going to be resolved as the FCC has moved the band where their operation is permitted from the 27 MHz CB band to 49 MHz, where such units will be competing only with themselves. The phase-out is very slow, and manufacturers can continue to make 27 MHz rigs for unlicensed use until March 1977, dealers can continue to sell them until March, 1978, and users can still operate them legally until March 1983.

However, since the "new generation" 49 MHz low power transceivers should provide their users with better range and performance than was ever possible on 27 MHz, units for the new band—still being developed as we go to press—will be a much better choice when they become available. If your need is for effective portable operation over short ranges (up to a mile or two in the open), this band should be for you! Interestingly enough, the rules for the new 49 MHz low power communications band even permit you to build your own transmitter, with the limitations that total power input be no more than 100 milliwatts and the antenna not exceed one meter in length and be mounted on the transmitter.

Pocket scanning receivers are also available in sizes little larger than a cigarette package. The RCA 16S100 is a 4-channel scanner that can monitor a combination of VHF and UHF frequencies.

LOW FREQUENCY EXPERIMENTAL OPERATION

Part 15 of the Rules also permits an individual to operate a transmitter on two low frequency bands without benefit of a license. One of these is the AM broadcast band, 510-1600 kHz, where a maximum power input to the final amplifier must not excede 100 milliwatts and the antenna plus transmission line is limited to 10 feet.

SBE's Opti/Scan sets frequencies by means of a slip-in punch card and permits the user to change from one group of 10 preselected frequencies to another instantly.

Older receivers offer a fine value for the shortwave listener. This 40-year-old National HRO may be scratched and dirty, but it's still an excellent SWL receiver. This example sold, complete with extra coils, for $35 at a recent hamfest.

The other band is in the "long wave" portion of the spectrum, 160-190 kHz. In this band a 50-foot antenna and one-watt input is permitted.

Despite these rather severe limitations some intrepid experimenters have achieved very remarkable results on these frequencies. By carefully picking frequencies away from local broadcast stations, some experimenters have made two-way contacts on the high end of the broadcast band at distances of a dozen miles or more. Experimenters operating on the long wave band have done even better, with 25- to 50-mile contacts common and a few confirmed reception reports at several hundred miles!

Though some of the workers on these frequencies are hams, many are simply people interested in radio and what it can do—shortwave listeners with a bent for experimentation. One very devoted long wave experimenter, Ken Cornell of Westfield, New Jersey, has even written a book—*The Low & Medium Frequency Scrapbook*—on the subject. If you'd like to know more about it, you should pick up a copy. It's $4.75 a copy and is distributed by *Ham Radio Magazine,* Greenville, New Hampshire 03048.

appendix A

MANUFACTURER'S DIRECTORY

Able Mfr. Co., Inc.
12820 So. Seventh St.
Grandview, MO 64030
(816) 966-8484
CB antennas; alarm system; radio brackets; antenna mounts.

Acoustic Fiber Sound Systems, Inc.
2831 No. Webster Ave.
Indianapolis, IN 46219
(317) 545-2481
CB communications speakers.

Airequipt, Inc.
1301 Brummel Avenue
Elk Grove, Illinois 60007
(312) 569-3556
CB antennas

Amco Electronics
18812 Bryant St.
Northridge, CA 91324
(213) 886-3864
CB switching panel.

American Electronics, Inc.
91 No. McKinley
Greenwood, IN 46142
(317) 888-7265
CB Radios and Accessories.

Anixter-Mark
5439 W. Fargo Ave.
Skokie, IL 60076
(312) 675-1500
Antennas and Antenna accessories; for CB, Marine and amatuer radio.

Antenna, Inc.
23850 Commerce Park Rd.
Cleveland, OH 44122
(216) 464-7075
CB antennas and antenna accessories.

The Antenna Specialists Co.
12435 Euclid Ave.
Cleveland, OH 44106
(216) 791-7878
CB, amatuer and marine antennas and accessories.

Apollo Products
532A East Edna Pl.
Covina, CA 91723
(213) 966-3031
Speakers and mounting brackets, plugs; jacks; cables; DC filter chokes; fuses; fuse holders; switches, solder irons; adapters.

Arista Enterprises Inc.
35 Hoffman Ave.
Hauppauge, NY 11787
(516) 234-7000
CB Antennas and mounts, microphones; meters, lock mount brackets; power supplies; noise filters.

Astatic Corp.
Corner Harbor & Jackson Streets
Conneaut, OH 44030
(216) 593-1111
CB microphones.

Audiovox Corp.
150 Marcu Blvd.
Hauppauge, NY 11787
(516) 231-7750
CB transceivers; test meters; antennas; and converters;

Automatic Radio
Two Main St.
Melrose, MA 02176
(617) 321-2300
CB radios and accessories.

Avanti R & D, Inc.
340 Stewart Ave.
Addison, IL 60101
(312) 543-9350
CB antennas and accessories.

B&B Import-Export, Inc.
185 Park St.
Troy, MI 48084
(313) 585-8400
CB transceivers; antennas; and accessories.

B&K-Precision
Div. Dynascan Corp.
1801 Belle Plaine
Chicago, IL 60613
(312) 327-7270
CB test instruments and power supplies.

Beltek Corp. of America
17910 So. LaSalle Ave.
Gardena, CA 90248
(213) 532-0254
CB Radios.

Bering Sales, Inc.
2777 "E" Irving Blvd.
P.O. Box 10962
Dallas, TX 75207
(214) 631-7470
CB radio mounts.

Blazer Communications, Inc.
34 Mildred Dr.
Fort Myers, FL 33901
(813) 936-8581
CB antennas and mounts.

Blue Streak Antenna
Div. CPD Industries
2100 E. Wilshire Ave.
Santa Ana, CA 92705
CB Antennas.

Boman Industries
9300 Hall Rd.
Downey, CA 90241
(213) 869-4041
CB Radios.

Bomar Crystal Company
201 Blackford Ave.
Middlesex, NJ 08846
(201) 356-7787
CB Crystals.

Breaker Corp.
1101 Great SW Pkwy.
Arlington, TX 76011
(817) 461-5061
CB antennas and accessories.

Browning Laboratories, Inc.
1269 Union Ave.
Laconia, NH 03246
(603) 524-5454
CB Radios.

C.P.D. Industries, Inc.
2100 E. Wilshire Ave.
Santa Ana, CA 92705
(714) 542-7228
CB antennas and mounts.

Cobra Communications Products Group, Dynascan Corp.
1801 W. Belle Plaine
Chicago, IL 60613
(312) 327-7270
CB radios & accessories.

Colt Communications
Div. of Directional International, Ltd.
5725 N. Central Avenue
Chicago, IL 60646
(312) 763-8440
CB radios

COMM Industries, Inc.
One Gateway Center
Newton, MA 02158
(617) 332-2266
CB Speakers.

Commando Communications Corp.
PO Box 11071
Chattanooga, TN 37401
(615) 756-8981
CB transceivers; radios, antennas; & accessories.

Communication Products Mfr., Inc.
9516 E. Montgomery, Suite 17
Spokane, WA 99206
(509) 928-3313
CB antennas & mounting hardware.

Communications Power, Inc.
2407 Charleston Rd.
Mountain View, CA 94043
(415) 965-2623
CB radios & accessories.

Consolidated Mfr. Co., Inc.
PO Box 6207-A
Birmingham, AL 35217
(205) 853-8825
Masts & towers.

Cornell-Dubilier Electronics
150 Avenue L
Newark, NJ 07101
CB noise filters; CB antennas; rotors & power supplies.

Courier Division
Fanon/Courier Corp.
990 So. Fair Oaks Ave.
Pasadena, CA 91105
(213) 799-9164
CB Radios.

Craig Corp.
921 W. Artesia Blvd.
Compton, CA 90220
Equip. Div.
(213) 537-1233
CB radios, antennas & accessories; scanners.

Cush Craft Corp.
621 Hayward St.
Manchester, NH 03103
(603) 627-7877
CB & amateur antennas.

Cushman Electronics, Inc.
830 Stewart Dr.
Sunnyvale, CA 94086
(408) 739-6760
Communications test instruments;

Manufacturer's Directory

Digital Sport Systems
7th & Elm St.
West Liberty, IO 52776
(319) 627-4211/4212
CB and amateur test instruments.

R. L. Drake Co.
540 Richard St.
Miamisburg, OH 45342
(513) 866-3621
SWL receivers, amateur equipment & accessories.

Dyn Electronics, Inc.
3095 NW 77th Ave.
Miami, FL 33122
(305) 592-6710
CB Radios.

EICO Electronic Instrument Co., Inc.
283 Malta St.
Brooklyn, New York 11207
(212) 272-1100
CB radios and test equipment.

Electric Co.
Div. Masco Corp.
300 South on E. County Line Rd.
Cumberland, IN 46229
(317) 894-3629
Scanner receivers.

Electricon Corp.
311 So. Park Dr.
Saint Marys, OH 45885
(419) 394-3363
CB antennas & power supplies.

Electronics Group, Inc.
2204 Foster Ave.
Wheeling, IL 60090
(312) 392-6030
CB radios, converters and test equipment.

Empire Machines & Systems, Inc.
Shore Rd.
Glenwood Landing, NY 11547
(516) 671-8200
CB security systems.

Everhardt Mfr. Co.
212 East St.
Hobart, IN 46342
(219) 942-3587
CB antennas & accessories.

Fanon Div.
Fanon/Courier Corp.
990 So. Fair Oaks Ave.
Pasadena, CA 91105
(213) 799-9164
CB radio scanners.

Far Eastern Research Laboratories, Inc.
8749 Shirley Ave.
Northridge, CA 91324
(213) 993-9101
CB radios & antennas.

Fieldmaster Radio Corp.
21212 Van Owen St.
Canoga Park, CA 91303
(213) 347-6810
CB radios and antennas; power supplies; scanners; marine and amateur equipment & accessories.

Fried Trading Co., Inc.
167 Clymer St.
Brooklyn, NY 11211
(212) 387-1157
CB radios, mounts and brackets; scanners; antennas.

Arthur Fulmer Communications
Box 177
Memphis, TN 38101
CB radios and antennas.

GC Electronics
400 So. Wyman
Rockford, IL 61101
(815) 968-9661
CB radios, accessories and antennas.

GEC, Inc.
4014 Nasa Road, 1
Seabrook, TX 77586
(713) 334-5343
CB instruments & accessories.

Gemtronics
356 South Blvd.
Lake City, SC 29560
(803) 394-3565
CB and marine radios; scanning monitors.

General Aviation Electronics, Inc.
4141 Kingman Dr.
Indianapolis, IN 46226
(317) 546-1111
Marine, amateur and aircraft radios: Mobiline 1; Mobiline ii: Mobiline III; marine master 25 W/a; marine master 202; marine mate 100; marine mate 1, GHT-1 hand held transceiver; GHT-6 hand-held transceiver; PSI-10 power supply; A/100-360 VHF-AM aircraft transceiver; GSB-1000 single side band transceiver.

Gold Line Connector, Inc.
25 Van Zant St.
East Norwalk, CT 06855
(203) 838-6551
CB accessories.

Handic — USA, Inc.
14560 NW 60 Ave.
Miami Lakes, FL 33014
(305) 558-1522
CB radios & accessories.

Herald Electronics
American Electronic Parts Corp.
6611 N. Lincoln Ave.
Chicago, IL 60645
(312) 675-1100
CB accessories and test equipment.

Hide-Away Antenna Mount
2828 Telephone Rd.
Houston, TX 77023
(712) 926-1723
CB and Amateur Antenna Mount.

Hitachi Sales Corp. of America
401 W. Artesia Blvd.
Compton, CA 90220
(213) 537-8383
CB Radios.

Hufco
PO Box 357
Provo, UT 84601
(801) 375-8566
CB accessories and test equipment: Digital counter line; CB vox; CB timer; digital readout attachment; CB/ham handbook.

Hy-Gain Electronics Corp.
8601 NE Highway 6
Lincoln, NB 68505
(402) 464-9151
CB radios, antennas & accessories.

I.A. Sales Co. of California, Inc.
766 Lakefield Rd., Suite H
Westlake Village, CA 91361
(805) 497-3966
CB Radios.

I.M.T. Corporation of Texas
PO Box 58147
Dallas, TX 75258
CB radios, antennas & accessories.

Itera Ltd.
1535 Broad St.
No. Bellmore, NY 11710
(516) 785-6480
CB power supplies & accessories.

JFD Electronics Corp.
Pine Tree Rd.
Oxford, NC 27565
(919) 693-3141
CB antennas & accessories.

J.I.L. Corporation of American, Inc.
1000 E. Del Amo Blvd.
Carson, CA 90746
(213) 537-7310
CB Radios.

Jasco Products Co.
PO Box 1513 — 4701 North Stiles
(405) 525-6333
CB accessories & antennas.

Ray Jefferson
Main & Cotton Streets
Philadelphia, PA 19127
(215) 487-2800
CB and marine radios; CB antennas.

Jemco CB Distributors, Inc.
2660 Electronics Lane
Dallas, TX 75220
Nelson Roberts
(214) 357-8386
CB Accessories.

E.F. Johnson Co.
299 10th Ave., SW
Waseca, MN 56093
(507) 835-2050
CB radios & scanners.

Kar Dol Corp.
1450 Dell Ave.
Campbell, CA 95008
(408) 378-6600
CB antennas & mounts.

Kraco Enterprises, Inc.
2411 No. Santa Fe Ave.
Compton, CA 90224
(213) 774-2550 or 639-0666
CB radios; antennas & accessories.

Kris, Inc.
N144 W5660 Pioneer Road
Cedarburg, WI 53012
(414) 375-1000
CB radios, antennas & accessories.

Kustom Kreations, Inc.
19316 Londelius St.
Northridge, CA 91324
(213) 886-8383
CB Accessories.

Kyodo Tsusho Corp.
10435 Lindley #126
Northridge, CA 91324
(213) 368-2215
CBs.

Lafayette Electronics Sales, Inc.
111 Jericho Turnpike
Syosset, NY 11791
(516) 921-7700
CB radios; antennas & accessories.

Lake Electronics
1948 — E Lehigh Ave.
Glenview, IL 60025
(312) 729-6767
CB radios; antennas and accessories; scanners.

Leader Instruments Corp.
151 Dupont St.
Plainview, NY 11803
(516) 822-9300
CB test instruments.

Lindsay Specialty Products Ltd.
50 Mary St.
Lindsay, Ont., Canada K9V 4S7
(705) 32402196
CB antennas.

MCM Mfr. Co., Inc.
6200 So. Freeway
Fort Worth, TX 76134
(817) 293-7533
CB antennas.

Magnadyne Corp.
PO Box 5365 — 20545 So. Belshaw Ave.
Carson, CA 90749
(213) 639-2200
CB antennas & accessories.

Manfred Enterprises
PO Box 29722
Dallas, TX 75229
(214) 241-3176
CB mounts.

Magitran Company
Div. ERA Acoustics Corp.
Moonachie, New Jersey 07074
(201) 641-3650
Alarm systems.

Metro Sound
11144 Weddington St.
No. Hollywood, CA 91601
CB radios; antennas and lock mounts.

Midland Int'l. Corp. Communication Div.
1909 Vernon St.
No. Kansas City, MO 64116
(816) 474-5080
CB radios; marine radios; scanning monitors; amateur radio equipment; antennas; test equipment & accessories.

Mid-West Electronics, Inc.
228 Fassett St.
Toledo, OH 43605
(419) 691-6111
CB antennas and mounts.

Miltronics, Inc.
12025 Manchester
St. Louis, MO 63131
Teargas Protector for mobile installations.

Misco Sales Co.
1235 Homestead St., Suite D
LaGrange Park, IL 60625
CB antennas & accessories.

Motorola Automotive Products Div.
1299 E. Algonquin Rd.
Schaumburg, IL 60172
(312) 576-2777
CB and marine radios & antennas.

Monitor Crystal Service, Inc.
PO Box 237
Watseka, IL 60970
(815) 432-5296
Crystals for scanning monitors and CB radios; amateur and CB radio accessories.

Mosley Electronics, Inc.
4610 No. Lindbergh Blvd.
Bridgeton, MO 63044
(314) 731-3036
CB and amateur antennas.

M-Tron Industries, Inc.
100 Douglas St.
Jankton, SD 57078
(605) 665-9321
CB and scanner crystals.

Mura Corp.
50 So. Service Road
Jericho, NY 11753
(516) 334-2700
CB antennas and test equipment.

New-Tronics Corp.
15800 Commerce Park Dr.
Brook Park, OH 44142
CB and Amateur antennas & mounts.

Numark Electronics Corp.
503 Raritan Center
Edison, NJ 08817
(201) 225-3222
CB radios, accessories; and antennas.

Nuvox Electronics Corp.
2 West 20th St.
New York, NY 10011
(212) 243-2110
CB radios and antennas.

On Guard Corporation of America
350 Gotham Pkwy.
Carlstadt, NJ 07072
(201) 939-5822
Vehicle alarms and CB lock mounts.

PAL Electronics Company
Div. of Fire Communications Corp.
2962 West Weldon
Phoenix, AZ 85017
(602) 264-0214
CB radios, antennas and test equipment.

Palomar Electronics
665 Opper St.
Escondido, CA 92025
(714) 746-2666
Amateur and CB radios & accessories.

Pathcom Inc., Pace Two-Way Radio Products
24049 So. Frampton Ave.
Harbor City, CA 90710
(213) 325-1290
CB radios; scanning monitors and marine radios.

Pearce-Simpson, Div. of Gladding Corp.
4701 NW 77th Ave.
Miami, FL 33166
(305) 592-5550

Platt Luggage, Inc.
2301 So. Prairie Ave.
Chicago, IL 60616
(312) 225-6670
CB cases.

Pride Electronics
7322 Convoy Court
San Diego, CA 92111
(714) 292-0766
CB test equipment.

RCA Distr. & Special Products Div.
Cherry Hill Offices, 206-2
Camden, NJ 08101
CB radios and scanning monitors.

Radio Shack
Box 1052
Fort Worth, TX 76101
CB radios, antenna & accessories.

Raider Corp.
1138 E. Big Beaver Rd.
Troy, MI 48084
(313) 689-9480
CB antennas.

Recoton Corp.
46 — 23 Crane St.
Long Island City, NY 11101
(212) 392-6442
CB mounts and accessories.

Regency Electronics, Inc.
7707 Records St.
Indianapolis, IN 46226
CB, Amateur and marine radios, scanner receivers.

Robyn Intl., Inc.
PO Box 478
Rockford, MI 49341
CB radios; scanners.

Royce Electronics Corp.
1746 Levee Rd.
No. Kansas City, MO 64116
(816) 842-7505
CB and marine radios; CB accessories; CB and marine antennas.

Russell Industries, Inc.
96 Station Plaza
Lynbrook, NY 11563
(516) 887-9000
CB antennas & accessories.

Rystl electronics Corp.
328 NW 170th St.
No. Miami Beach, FL 33169
(305) 652-3838
CB radios & antennas.

S & A Electronics
Div. of The Scott & Fetzer Co.
202 W. Florence St.
Toledo, OH 43605
(419) 693-0528
CB antennas.

SBE, Inc.
220 Airport Blvd.
Watsonville, CA 95076
(408) 722-4177
CB and marine transceivers; scanning monitors; CB radios.

SST South Shore Trading Corp.
422 S. Franklin St.
Hempstead, NY 11550
(516) 486-7100
CB mounts and accessories.

Saxton Products
215 No. Route 303
Congers, NY 10920
(914) 268-6846 or (815) 962-5041
CB antennas & accessories.

Shakespeare Co.
PO Box 246
Columbia, SC 29202
(803) 779-5800
CB and marine radios; marine antennas; & accessories.

Sharp Electronics Corp.
10 Keystone Place
Paramus, NJ 07652
(201) 265-5600
CB radios and antennas.

Shur-Lok Mfr. Co., Inc.
413 North Main
Hutchins, TX 75141
(214) 225-2009
CB radio locking mounts.

Sidewinder Intl. U.S.A., Inc.
3570 7th St.
Wayland, MI 49348
(616) 792-2205
CB radios and antennas; test equipment and accessories.

Siltronix
A Div. of Cubic Corp.
330 Via El Centro
Oceanside, CA 92054
(714) 757-8860
CB radios & accessories.

Sonar Radio Corporation
3918 No. 29th Avenue
Hollywood, FL 33020
(305) 925-3400
CB radios.

South River Metal Products Co., Inc.
377-379 Turnpike Rd.
South River, NJ 08882
(201) 254-5252
Antenna towers; mounts and brackets; antenna mounting accessories.

SouthCom, Inc.
PO Box 11212
Fort Worth, TX 76109
(817) 293-6590
Mobile antenna mounts.

Sparkomatic Corp.
Milford, PA 18337
(717) 296-6444
CB radios, converts and antennas.

Specialties, Inc.
393 South Franklin St.
Hempstead, New York 11550
(516) 292-9500
CB speakers.

Standard Communications Corp.
PO Box 92151
Los Angeles, CA 90009
(213) 532-5300
CB, marine and amateur radios.

Manufacturer's Directory

Surveyor Mfr. Corp.
7 Electronics Court
Madison Heights, MI 48071
(313) 544-9110
CB radios; antennas & accessories; scanners.

T.E.S.T., Inc.
19428 Londelius St.
Northridge, CA 91324
(213) 349-2403
CB converter; noise eliminator.

TRS Intl. Ltd.
4825 N. Scott St., Suite 214
Schiller Pk., IL 60176
(312) 678-5227
CB radios; antennas; and accessories.

Taylor Radio Co., Inc.
3329 Commerce Dr.
Augusta, GA 30904
(404) 736-3336
CB antennas & accessories.

Teaberry Electronics Corp.
6330 Castleplace Dr.
Indianapolis, IN 46250
(317) 545-1201
CB radios.

Telco Products Corp.
44 Sea Cliff Ave.
Glen Cove, NY 11542
(516) 759-0300
CB accessories and test equipment.

Tenna Corp.
19201 Cranwood Pkwy.
Warrensville Heights, OH 44128
(216) 475-1400
CB converters; and antennas.

Time Mfr. Co.
PO Box 116
Napoleon, MO 64074
(816) 934-8114/8115
CB antennas and mounting brackets.

Tram/Diamond Corp.
PO Box 187, Lower Bay Rd.
Winnisquam, NH 03289
(603) 524-0622
CB radios.

Trans-Comm Mfr. Co., Inc.
8885 Bond St.
Overland Park, KS 66214
(913) 888-9115
CB radios and accessories.

Tran-Sonic Industries, Inc.
12 Fairview Terrace
Paramus, NJ 07652
(201) 845-0370
CB radios and antennas.

Tri-Ex Tower Corp.
7182 Rasmussen Ave.
Visalia, CA 93277
(209) 732-8383
Towers

True Temper Corp.
PO Box 1050
Anderson, SC 29621
(803) 224-0271
CB antennas.

Turner
Div. of Conrac Corp.
716 Oakland Road NE
Cedar Rapids, IO 52402
(319) 365-0421
CB microphones and antennas.

Ultratec/Workman Electronics
PO Box 3828
Sarasota, FL 33578
(813) 371-4242
CB accessories.

Unarco-Rohn
Div. of Unarco Inds., Inc.
6718 West Plank Rd.—PO Box 2000
Peoria, IL 61601
(309) 697-4400
Towers and accessories.

Unimetrics, Inc.
123 Jericho Turnpike
Syosset, NY 11791
(516) 364-8100
CB and marine radios.

United States Crystal Corp.
3605 McCart St.
Fort Worth, TX 76110
(817) 921-3014
Crystals for CB radios and monitor receivers.

Universal Security Instruments, Inc.
2829 Potee St.
Baltimore, MD 21225
(301) 355-9000
CB accessories.

Valor Enterprises
185 W. Hamilton
No. Milton, OH 45318
(513) 698-4194
CB antennas and accessories.

Van Ordt, Inc.
10875 So. Grand Ave.
Ontario, CA 91762
(714) 628-4791
CB radio and antenna mounting devices; antennas.

Vanco-Chicago, Inc.
970 North Shore Drive
Lake Bluff, IL 60044
(312) 295-1050
CB antennas, instruments and accessories.

Vector, Inc.
23824 Hawthorne Blvd.
Torrance, CA 90505
CB radios.

Vandetta Corp.
431 Monte Vista Dr.
Dallas TX 75223
(214) 321-1864
CB antennas; and mounting brackets.

Wilson Electronics
4288 So. Polaris Ave.
Las Vegas, NV 89103
(702) 739-1931
CB and amateur antennas; amateur radios.

Windsor Industries, Inc.
10 Hub Drive
Melville, Long Island, NY 11746
(516) 694-1400
CB radios

Zodiac Communications Corp.
626 Chrysler Bldg.
New York, NY 10017
(212) 697-9585
CB radios; accessories and test equipment.

appendix B

FCC RULES AND REGULATIONS

Contents—Part 95

Subpart A—General

Sec.
- 95.1 Basis and purpose.
- 95.3 Definitions.
- 95.5 Policy governing the assignment of frequencies.
- 95.6 Types of operation authorized.
- 95.7 General citizenship requirements.

Subpart B—Applications and Licenses

- 95.11 Station authorization required.
- 95.13 Eligibility for station license.
- 95.14 Mailing address furnished by licensee.
- 95.15 Filing of applications.
- 95.17 Who may sign applications.
- 95.19 Standard forms to be used.
- 95.25 Amendment or dismissal of application.
- 95.27 Transfer of license prohibited.
- 95.29 Defective applications.
- 95.31 Partial grant.
- 95.33 License term.
- 95.35 Changes in transmitters and authorized stations.
- 95.37 Limitations on antenna structures.

Subpart C—Technical Regulations

- 95.41 Frequencies available.
- 95.42 Special Provisions.
- 95.43 Transmitter power.
- 95.44 External radio frequency power amplifiers prohibited.
- 95.45 Frequency tolerance.

Sec.
- 95.47 Types of emission.
- 95.49 Emission limitations.
- 95.51 Modulation requirements.
- 95.53 Compliance with technical requirements.
- 95.55 Acceptability of transmitters for licensing.
- 95.57 Procedure for type acceptance of equipment.
- 95.58 Additional requirements for type acceptance.
- 95.59 Submission of noncrystal controlled Class C station transmitters for type approval.
- 95.61 Type approval of receiver-transmitter combinations.
- 95.63 Minimum equipment specifications.
- 95.65 Test procedure.
- 95.67 Certificate of type approval.

Subpart D—Station Operating Requirements

- 95.81 Permissible communications.
- 95.83 Prohibited communications.
- 95.85 Emergency and assistance to motorist use.
- 95.87 Operation by, or on behalf of, persons other than the licensee.
- 95.89 Telephone answering services.
- 95.91 Duration of transmissions.
- 95.93 Tests and adjustments.
- 95.95 Station identification.
- 95.97 Operator license requirements.
- 95.101 Posting station license and transmitter identification cards or plates.

Sec.
- 95.103 Inspection of stations and station records.
- 95.105 Current copy of rules required.
- 95.107 Inspection and maintenance of tower marking and lighting, and associated control equipment.
- 95.111 Recording of tower light inspections.
- 95.113 Answers to notices of violations.
- 95.115 False signals.
- 95.117 Station location.
- 95.119 Control points, dispatch points, and remote control.
- 95.121 Civil defense communications.

Subpart E—Operation of Citizens Radio Stations in the United States by Canadians

- 95.131 Basis, purpose and scope.
- 95.133 Permit required.
- 95.135 Application for permit.
- 95.137 Issuance of permit.
- 95.139 Modification or cancellation of permit.
- 95.141 Possession of permit.
- 95.143 Knowledge of rules required.
- 95.145 Operating conditions.
- 95.147 Station identification.

AUTHORITY: §§95.1 to 95.147 issued under secs. 4, 303, 48 Stat. 1066, 1082, as amended; 47 U.S.C. 154, 303. Interpret or apply 48 Stat. 1064—1068, 1081—1105, as amended; 47 U.S.C. Sub-chap. I, III—VI.

FEDERAL COMMUNICATIONS COMMISSION RULES AND REGULATIONS
PART 95/CITIZENS RADIO SERVICE

Revised October 29, 1976

(NOTE: Rules most significant to Class D—27 MHz—CB operators are **bold face** type)

SUBPART A—GENERAL

95.1 Basis and purpose.

The rules and regulations set forth in this part are issued pursuant to the provisions of Title III of the Communications Act of 1934, as amended, which vests authority in the Federal Communications Commission to regulate radio transmissions and to issue licenses for radio stations. **These rules are designed to provide for private short-distance radiocommunications service for the business or personal activities of licensees, for radio signaling, for the control of remote objects or devices by means of radio; all to the extent that these uses are not specifically prohibited in this part.** They also provide for procedures whereby manufacturers of radio equipment to be used or operated in the Citizens Radio Service may obtain type acceptance and/or type approval of such equipment as may be appropriate.

95.3 Definitions.

For the purpose of this part, the following definitions shall be applicable. For other definitions, refer to Part 2 of this chapter.

(a) Definitions of services.

Citizens Radio Service. A radio-communications service of fixed, land, and mobile stations intended for short-distance personal or business radiocommunications, radio signaling, and control of remote objects or devices by radio; all to the extent that these uses are not specifically prohibited in this part.

Fixed Service. A service of radio-communication between specified fixed points.

Mobile service. A service of radio-communication between mobile and land stations or between mobile stations.

(b) Definitions of stations.

Base Station. A land station in the land mobile service carrying on a service with land mobile stations.

Class A station. A station in the Citizens Radio Service licensed to be operated on an assigned frequency in the 460—470 MHz band with a transmitter output power of not more than 50 watts.

Class B station. (All operations terminated as of November 1, 1971.)

Class C station. A station in the Citizens Radio Service licensed to be operated on an authorized frequency in the 26.96—27.23 MHz band, or on the frequency 27.255 MHz, for the control of remote objects or devices by radio, or for the remote actuation of devices which are used solely as a means of attracting attention, or on an authorized frequency in the 72—76 MHz band for the radio control of models used for hobby purposes only.

Class D station. A station in the Citizens Radio Service licensed to be operated for radiotelephony, only, on authorized frequencies in the 26.96 MHz to 27.41 MHz band.

Fixed station. A station in the fixed service.

Land station. A station in the mobile service not intended for operation while in motion. (Of the various types of land stations, only the base station is pertinent to this part.)

Mobile station. A station in the mobile service intended to be used while in motion or during halts at unspecified points. (For the purposes of this part, the term includes hand-carried and pack-carried units.)

(c) Miscellaneous definitions.

Antenna structures. The term "antenna structures" includes the radiating system its supporting structures and any appurtenances mounted thereon.

Assigned frequency. The frequency appearing on a station authorization from which the carrier frequency may deviate by an amount not to exceed that permitted by the frequency tolerance.

Authorized bandwidth. The maximum permissible bandwidth for the particular emission used. This shall be the occupied bandwidth or necessary bandwidth, whichever is greater.

Carrier power. The average power at the output terminals of a transmitter (other than a transmitter having a suppressed, reduced or controlled carrier) during one radio frequency cycle under conditions of no modulation.

Control point. A control point is an operating position which is under the control and supervision of the licensee, at which a person immediately responsible for the proper operation of the transmitter is stationed, and at which adequate means are available to aurally monitor all transmissions and to render the transmitter inoperative.

Dispatch point. A dispatch point is any position from which messages may be transmitted under the supervision of the person at a control point.

Double sideband emission. An emission in which both upper and lower sidebands resulting from the modulation of a particular carrier are transmitted. The carrier, or a portion thereof, also may be present in the emission.

External radio frequency power amplifiers. As defined in 2.815 (a) and as used in this part, an external radio frequency power amplifier is any device which, (1) when used in conjunction with a radio transmitter as a signal source is capable of amplification of that signal, and (2) is not an integral part of a radio transmitter as manufactured.

Harmful interference. Any emission, radiation or induction which endangers the functioning of a radionavigation service or other safety service or seriously degrades, obstructs or repeatedly interrupts a radiocommunication service operating in accordance with applicable laws, treaties, and regulations.

Man-made structure. Any construction other than a tower, mast or pole.

Mean power. The power at the output terminals of a transmitter during normal operation, averaged over a time sufficiently long compared with the period of the lowest frequency encountered in the modulation. A time of 1/10 second during which the mean power is greatest will be selected normally.

Necessary bandwidth. For a given class of emission, the minimum value of the occupied bandwidth sufficient to ensure the transmission of information at the rate and with the quality required for the system employed, under specified conditions. Emissions useful for the good functioning of the receiving equipment, as for example, the emission corresponding to the carrier of reduced carrier systems, shall be included in the necessary bandwidth.

Occupied bandwidth. The frequency bandwidth such that, below its lower and above its upper frequency limits, the mean powers radiated are each equal to 0.5% of the total mean power radiated by a given emission.

Omnidirectional antenna. An antenna designed so the maximum radiation in any horizontal direction is within 3 dB of the minimum radiation in any horizontal direction.

Peak envelope power. The average power at the output terminals of a transmitter during one radio frequency cycle at the highest crest of the modulation envelope, taken under conditions of normal operation.

Person. The term "person" includes an individual, partnership, association, joint-stock company, trust or corporation.

Remote control. The term "remote control" when applied to the use or operation of a citizens radio station means control of the transmitting equipment of that station from any place other than the location of the transmitting equipment, except that direct mechanical control or direct electrical control by wired connections of transmitting equipment from some other point on the same premises, craft or vehicle shall not be considered to be remote control.

Single sideband emission. An emission in which only one sideband is transmitted. The carrier, or a portion thereof, also may be present in the emission.

Station authorization. Any construction permit, license, or special temporary authorization issued by the Commission.

95.5 Policy governing the assignment of frequencies.

(a) The frequencies which may be assigned to Class A stations in the Citizens Radio Service, and the frequencies which are available for use by Class C or Class D stations are listed in Subpart C of this part. **Each frequency available for assignment to, or use by, stations in this service is available on a shared basis only, and will not be assigned for the exclusive use of any one applicant; however, the use of a particular frequency may be restricted to (or in) one or more specified geographical areas.**

(b) In no case will more than one frequency be assigned to Class A stations for the use of a single applicant in any given area until it has been demonstrated conclusively to the Commission that the assignment of an additional frequency is essential to the operation proposed.

(c) All applicants and licensees in this service shall cooperate in the selection and use of the frequencies assigned or authorized, in order to minimize interference and thereby obtain the most effective use of the authorized facilities.

(d) Simultaneous operation on more than one frequency in the 72—76 MHz band by a transmitter or transmitters of a single licensee is prohibited whenever such operation will cause harmful interference to the operation of other licensees in this service.

95.6 Types of operation authorized.

(a) Class A stations may be authorized as mobile stations, as base stations, as fixed stations, or as base or fixed stations to be operated at unspecified or temporary locations.

(b) Class C and Class D stations are authorized as mobile stations only; however, they may be operated at fixed locations in accordance with other provisions of this part.

95.7 General citizenship requirements.

A station license shall not be granted to or held

by a foreign government or a representative thereof.
[95.7 revised eff. 2-5-75; VI(75)-1]

SUBPART A—GENERAL
SUBPART B—APPLICATIONS AND LICENSES

95.11 Station authorization required.

No radio station shall be operated in the Citizens Radio Service except under and in accordance with an authorization granted by the Federal Communications Commission.

95.13 Eligibility for station license.

(a) Subject to the general restrictions of 95.7, any person is eligible to hold an authorization to operate a station in the Citizens Radio Service: *Provided,* That if an applicant for a Class A or Class D station authorization is an individual or partnership, such individual or each partner is eighteen or more years of age; or if an applicant for a Class C station authorization is an individual or partnership, such individual or each partner is twelve or more years of age. An unincorporated association, when licensed under the provisions of this paragraph, may upon specific prior approval of the Commission provide radiocommunications for its members.

NOTE: While the basis of eligibility in this service includes any state, territorial, or local governmental entity, or any agency operating by the authority of such governmental entity, including any duly authorized state, territorial, or local civil defense agency, it should be noted that the frequencies available to stations in this service are shared without distinction between all licensees and that no protection is afforded to the communications of any station in this service from interference which may be caused by the authorized operation of other licensed stations.

(b) [Reserved]

(c) No person shall hold more than one Class C and one Class D station license.

95.14 Mailing address furnished by licensee.

Except for applications submitted by Canadian citizens pursuant to agreement between the United States and Canada (TIAS No. 2508 and No. 6931), each application shall set forth and each licensee shall furnish the Commission with an address in the United States to be used by the Commission in serving documents or directing correspondence to that licensee. Unless any licensee advises the Commission to the contrary, the address contained in the licensee's most recent application will be used by the Commission for this purpose.

[95.14 amended eff. 12-17-75; VI (75)-2]

95.15 Filing of applications.

(a) To assure that necessary information is supplied in a consistent manner by all persons, standard forms are prescribed for use in connection with the majority of applications and reports submitted for Commission consideration. Standard numbered forms applicable to the Citizens Radio Service are discussed in 95.19 and may be obtained from the Washington, D.C., 20554, office of the Commission, or from any of its engineering field offices.

(b) All formal applications for Class C or Class D new, modified, or renewal station authorizations shall be submitted to the Commission's office, Box 1010, Gettysburg, Pa. 17325. An application for a temporary permit shall be made by completing and making certifications required by FCC Form 555-B. Applications for Class A station authorizations, applications for consent to transfer of control of a corporation holding any citizens radio station authorization, requests for special temporary authority or other special requests, and correspondence relating to an application for any class citizens radio station authorization shall be submitted to the Commission's Office at Washington, D.C. 20554, and should be directed to the attention of the Secretary. Applicants for Class A stations in the Chicago Regional Area, defined in 95.19, shall submit their application to the Commission's Chicago Regional Office. Applications involving Class A or Class D station equipment which is neither type approved nor crystal controlled, whether of commercial or home construction, shall be accompanied by supplemental data describing in detail the design and construction of the transmitter and methods employed in testing it to determine compliance with the technical requirements set fourth in Subpart Co of this part.

(c) Unless otherwise specified, an application shall be filed at least 60 days prior to the date on which it is desired that Commission action thereon be completed. In any case where the applicant has made timely and sufficient application for renewal of license, in accordance with the Commission's rules, no license with reference to any activity of a continuing nature shall expire until such application shall have been finally determined.

(d) A temporary permit may not be held by an applicant already holding a Class D station license.

(e) Failure on the part of the applicant to provide all the information required by the application form, or to supply the necessary exhibits or supplementary statements may constitute a defect in the application.

(f) Applicants proposing to construct a radio station on a site located on land under the jurisdiction of the U.S. Forest Service, U.S. Department of Agriculture, or the Bureau of Land Management, U.S. Department of the Interior, must supply the information and must follow the procedure prescribed by 1..70 of this chapter.

95.17 Who may sign applications.

(a) Except as provided in paragraph (b) of this section, applications, amendments thereto, and related statements of fact required by the Commission shall be personally signed by the applicant, if the applicant is an individual; by one of the partners, if the applicant is a partnership; by an officer, if the applicant is a corporation; or by a member who is an officer, if the applicant is an unincorporated association. Applications, amendments, and related statements of fact filed on behalf of eligible government entities, such as states and territories of the United States and political subdivisions thereof, the District of Columbia, and units of local government, including incorporated municipalities, shall be signed by such duly elected or appointed officials as may be competent to do so under the laws of the applicable jurisdiction.

(b) Applications, amendments thereto, and related statements of fact required by the Commission may be signed by the applicant's attorney in case of the applicant's physical disability or of his absence from the United States. The attorney shall in that event separately set forth the reason why the application is not signed by the applicant. In addition, if any matter is stated on the basis of the attorney's belief only (rather than his knowledge), he shall separately set forth his reasons for believing that such statements are true.

(c) Only the original of applications, amendments, or related statements of fact need be signed; copies may be conformed.

(d) Applications, amendments, and related statements of the fact need not be signed under oath. Willful false statements made therein, however, are punishable by fine and imprisonment. U.S. Code, Title 18, section 1001, and by appropriate administrative sanctions, including revocation of station license pursuant to section 312 (a) (1) of the Communications Act of 1934, as amended.

95.19 Standard forms to be used.

(a) *FCC Form 505, Application for Class C or D Station License in the Citizens Radio Service.* This form shall be used when:

(1) Application is made for a new Class C or Class D authorization. A separate application shall be submitted for each proposed class of station.

(2) Application is made for modification of any existing Class C or Class D station authorization in those cases where prior Commission approval of certain changes is required (see 95.35).

(3) Application is made for renewal of an existing Class C or Class D station authorization, or for reinstatement of such an expired authorization.

(b) FCC Form 555-B, *Temporary Permit, Class D Citizens Radio Station.* This form shall be used when application is made for a temporary permit.

(c) FCC Form 400, *Application for Radio Station Authorization in the Safety and Special Radio Services.* Except as provided in paragraph (d) of this section, this form shall be used when:

(1) Application is made for a new Class A base station or fixed station authorization. Separate applications shall be submitted for each proposed base or fixed station at different fixed locations; however, all equipment intended to be operated at a single fixed location is considered to be one station which may, if necessary, be classed as both a base station and a fixed station.

(2) Application is made for a new Class A station authorization for any required number of mobile units (including hand-carried and pack-carried units) to be operated as a group in a single radiocommunication system in a particular area. An application for Class A mobile station authorization may be combined with the application for a single Class A base station authorization when such mobile units are to be operated with that base station only.

(3) Application is made for station license of any Class A base station or fixed station upon completion of construction or installation in accordance with the terms and conditions set forth in any construction permit required to be issued for that station, or application for extension of time within which to construct such a station.

(4) Application is made for modification of any existing Class A station authorization in those cases where prior Commission approval of certain changes is required (see 95.35).

(5) Application is made for renewal of an existing Class A station authorization, or for reinstatement of such an expired authorization.

(6) Each applicant in the Safety and Special Radio Services (1) for modification of a station license involving a site change or a substantial increase in tower height or (2) for a license for a new station must, before commencing construction, supply the environmental information, where required, and must follow the procedure prescribed by Subpart I of Part 1 of this chapter (1.1301 through 1.1319 unless Commission action authorizing such construction would be a minor action within the meaning of Subpart I of Part 1.

(7) Application is made for an authorization for a new Class A base or fixed station to be operated at unspecified or temporary locations. When one or more individual transmitters are each intended to be operated as a base station or as a fixed station at unspecified or temporary locations for indeterminate periods, such transmitters may be considered to comprise a single station intended to be operated at temporary location. The appliction shall specify the general geographic area within which the operation will be confined. Sufficient data must be submitted to show the need for the proposed area of operation.

(d) *FCC Form 703, Application for Consent to Transfer of Control of Corporation Holding*

Construction Permit or Station License. This form shall be used when application is made for consent to transfer control of a corporation holding any citizens radio station authorization.

(e) Beginning April 1, 1972, FCC Form 425 shall be used in lieu of FCC Form 400, applicants for Class A stations located in the Chicago Regional Area defined to consist of the counties listed below:

ILLINOIS
1. Boone.
2. Bureau.
3. Carroll.
4. Champaign.
5. Christian
6. Clark.
7. Coles.
8. Cook.
9. Cumberland.
10. De Kalb.
11. De Witt.
12. Douglas.
13. Du Page.
14. Edgar.
15. Ford.
16. Fulton.
17. Grundy
18. Henry.
19. Iroquois.
20. Jo Daviess.
21. Kane.
22. Kankakee.
23. Kendall.
24. Knox.
25. Lake.
26. La Salle.
27. Lee.
28. Livingston.
29. Logan.
30. Macon.
31. Marshall.
32. Mason.
33. McHenry.
34. McLean.
35. Menard.
36. Mercer.
37. Moultrie.
38. Ogle.
39. Peoria.
40. Piatt.
41. Putnam.
42. Rock Island.
43. Sangamon.
44. Shelby.
45. Stark.
46. Stephenson.
47. Tazewell.
48. Vermillion.
49. Warren.
50. Whiteside.
51. Will.
52. Winnebago.
53. Woodford.

INDIANA
1. Adams.
2. Allen.
3. Benton.
4. Blackford.
5. Boone.
6. Carroll.
7. Cass.
8. Clay.
9. Clinton.
10. De Kalb.
11. Delaware.
12. Elkhart.
13. Fountain.
14. Fulton.
15. Grant
16. Hamilton.
17. Hancock.
18. Hendricks.
19. Henry.
20. Howard.
21. Huntington.
22. Jasper.
23. Jay.
24. Kosciusko.
25. Lake.
26. Lagrange.
27. La Porte.
28. Madison.
29. Marion.
30. Marshall.
31. Miami.
32. Montgomery.
33. Morgan.
34. Newton.
35. Noble.
36. Owen.
37. Parke.
38. Porter.
39. Pulaski.
40. Putnam.
41. Randolph.
42. St. Joseph.
43. Starke.
44. Steuben.
45. Tippecanoe.
46. Tipton.
47. Vermillion.
48. Vigo.
49. Wabash.
50. Warren.
51. Wells.
52. White.
53. Whitley.

IOWA
1. Cedar
2. Clinton
3. Dubuque
4. Jackson.
5. Jones
6. Muscatine.
7. Scott.

MICHIGAN
1. Allegan.
2. Barry.
3. Berrien.
4. Branch.
5. Calhoun.
6. Cass.
7. Clinton.
8. Eaton.
9. Hillsdale.
10. Ingham.
11. Ionia.
12. Jackson.
13. Kalamazoo.
14. Kent.
15. Lake.
16. Mason.
17. Mecosta.
18. Montcalm.
19. Muskegon.
20. Newaygo.
21. Oceana.
22. Ottawa.
23. St. Joesph.
24. Van Buren.

OHIO
1. Defiance.
2. Mercer.
3. Paulding.
4. Van Wert.
5. Williams.

WISCONSIN
1. Adams.
2. Brown.
3. Calumet.
4. Columbia.
5. Dane.
6. Dodge.
7. Door.
8. Fond du Lac.
9. Grant.
10. Green.
11. Green Lake.
12. Iowa.
13. Jefferson.
14. Juneau.
15. Kenosha.
16. Kewaunee.
17. Lafayette.
18. Manitowoc.
19. Marquette.
20. Milwaukee.
21. Outagamie.
22. Ozaukee.
23. Racine.
24. Richland.
25. Rock.
26. Sauk.
27. Sheboygan.
28. Walworth.
29. Washington.
30. Waukesha.
31. Waupaca.
32. Waushara.
33. Winnebago.

95.25 Amendment or dismissal of application.

(a) Any application may be amended upon request of the applicant as a matter of right prior to the time the application is granted or designated for hearing. Each amendment to an application shall be signed and submitted in the same manner and with the same number of copies as required for the original application.

(b) Any application may, upon written request signed by the applicant or his attorney, be dismissed without prejudice as a matter of right prior to the time the application is granted or designated for hearing.

95.27 Transfer of license prohibited.

A station authorization in the Citizens Radio Service may not be transferred or assigned. In lieu of such transfer or assignment, an application for new station authorization shall be filed in each case, and the previous authorization shall be forwarded to the Commission for cancellation.

95.29 Defective applications.

(a) If an applicant is requested by the Commission to file any documents or information not included in the prescribed application form, a failure to comply with such request will constitute a defect in the application.

(b) When an application is considered to be incomplete or defective, such application will be returned to the applicant, unless the Commission may otherwise direct. The reason for return of the applications will be indicated, and if appropriate, necessary additions or corrections will be suggested:

95.31 Partial grant.

Where the Commission, without a hearing, grants an application in part, or with any privileges, terms, or conditions other than those requested, the action of the Commission shall be considered as a grant of such application unless the applicant shall, within 30 days from the date on which such grant is made, or from its effective date if a later date is specified, file with the Commission a written rejection of the grant as made. Upon receipt of such rejection, the Commission will vacate its original action upon the application and, if appropriate, set the application for hearing.

95.33 License term.

Licenses for stations in the Citizens Radio Service will normally be issued for a term of 5 years from the date of original issuance, major modification, or renewal.

95.35 Changes in transmitters and authorized stations.

Authority for certain changes in transmitters and authorized stations must be obtained from the Commission before the changes are made, while other changes do not require prior Commission approval. The following paragraphs of this section describe the conditions under which prior Commission approval is or is not necessary.

(a) Proposed changes which will result in operation inconsistent with any of the terms of the current authorization require that an application for modification of license be submitted to the commission. Application for modification shall be submitted in the same manner as an application for a new station license, and the licensee shall forward his existing authorization to the Commission for cancellation immediately upon receipt of the superseding authorization. Any of the following changes to authorized stations may be made only upon approval by the Commission:

(1) Increase the overall number of transmitters authorized.

(2) Change the presently authorized location of a Class A fixed or base station or control point.

(3) Move, change the height of, or erect a Class A station antenna structure.

(4) Make any change in the type of emission or any increase in bandwidth of emission or power of a Class A station.

(5) Addition or deletion of control point(s) for an authorized transmitter of a Class A station.

(6) Change or increase the area of operation of a Class A mobile station or a Class A base or fixed station authorized to be operated at temporary locations.

(7) Change the operating frequency of a Class A station.

(b) When the name of a licensee is changed (without changes in the ownership, control, or corporate structure), or when the mailing address of the licensee is changed (without changing the authorized location of the base or fixed Class A station) a formal application for modification of the license is not required. However, the licensee shall notify the Commission promptly of these changes. The notice, which may be in letter form, shall contain the name and address of the licensee as they appear in the Commission's records, the new name and/or address, as the case may be, and the call signs and classes of all radio stations authorized to the licensee under this part. The notice concerning Class C or D radio stations shall be sent to Federal Communications Commission, Gettysburg, Pa. 17325, and a copy shall be maintained with the records of the station. The notice concerning Class A stations shall be sent to (1) Secretary, Federal Communications Commission, Washington, D.C. 20554, and (2) to Engineer in Charge of the Radio District in which the station is located, and a copy shall be maintained with the license of the station until a new license is issued.

(c) Proposed changes which will not depart from any of the terms of the outstanding authorization for the station may be made without prior Commission approval. Included in such changes is the substitution of transmitting equipment at any station, provided that the equipment employed is included in the Commission's "Radio Equipment List," and is listed as acceptable for use in the appropriate class of station in this service. Provided it is crystal-controlled and otherwise complies with the power, frequency tolerance, emission and modulation percentage limitations prescribed, non-type accepted equipment may be substituted at:

(1) Class C stations operated on frequencies in the 26.99—27.26 MHz band;

(2) Class D stations until November 22, 1974.

(d) Transmitting equipment type accepted for use in Class D stations shall not be modified by the user. Changes which are specifically prohibited include:

(1) Internal or external connection or addition of any part, device or accessory not included by the manufacturer with the transmitter for its type acceptance. This shall not prohibit the external

connection of antennas or antenna transmission lines, antenna switches, passive networks for coupling transmission lines or antennas to transmitters, or replacement of microphones.

(2) Modification in any way not specified by the transmitter manufacturer and not approved by the Commission.

(3) Replacement of any transmitter part by a part having different electrical characteristics and ratings from that replaced unless such part is specified as a replacement by the transmitter manufacturer.

(4) Substitution or addition of any transmitter oscillator crystal unless the crystal manufacturer or transmitter manufacturer has made an express determination that the crystal type, as installed in the specific transmitter type, will provide that transmitter type with the capability of operating within the frequency tolerance specified in Section 95.45 (a).

(5) Addition or substitution of any component, crystal or combination of crystals, or any other alteration to enable transmission on any frequency not authorized for use by the licensee.

(e) Only the manufacturer of the particular unit of equipment type accepted for use in Class D stations may make the permissive changes allowed under the provisions of Part 2 of this chapter for type acceptance. However, the manufacturer shall not make any of the following changes to the transmitter without prior written authorization from the Commission:

(1) Addition of any accessory or device not specified in the application for type acceptance and approved by the Commission in granting said type acceptance.

(2) Addition of any switch, control, or external connection.

(3) Modification to provide capability for an additional number of transmitting frequencies.

95.37 Limitations on antenna structures.

(a) Except as provided in paragraph (b) of this section, an antenna for a Class A station which exceeds the following height limitations may not be erected or used unless notice has been filed with both the FAA on FAA Form 7460—1 and with the Commission on Form 714 or on the license application form, and prior approval by the Commission has been obtained for:

(1) Any construction or alteration of more than 200 feet in height above ground level at its site (17.7 (a) of this chapter).

(2) Any construction or alteration of greater height than an imaginary surface extending outward and upward at one of the following slopes (17.7 (b) of this chapter):

95.37 Limitations on antenna structures.

(i) 100 to 1 for a horizontal distance of 20,000 feet from the nearest point of the nearest runway of each airport with at least one runway more than 3,200 feet in length, excluding heliports, and seaplane bases without specified boundaries, if that airport is either listed in the Airport Directory of the current Airman's Information Manual or is operated by a Federal military agency.

(ii) 50 to 1 for a horizontal distance of 10,000 feet from the nearest point of the nearest runway of each airport with its longest runway no more than 3,200 feet in length, excluding heliports, and seaplane bases without specified boundaries, if that airport is either listed in the Airport Directory or is operated by a Federal military agency.

(iii) 25 to 1 for a horizontal distance of 5,000 feet from the nearest point of the nearest landing and takeoff area of each heliport listed in the Airport Directory or operated by a Federal military agency.

(3) Any construction or alteration on any airport listed in the Airport Directory of the current Airman's Information Manual (17.7 (c) of this chapter).

(b) A notification to the Federal Aviation Administration is not required for any of the following construction or alteration of Class A station antenna structures.

(1) Any object that would be shielded by existing structures of a permanent and substantial character or by natural terrain or topographic features of equal or greater height, and would be located in the congested area of a city, town, or settlement where it is evident beyond all reasonable doubt that the structure so shielded will not adversely affect safety in air navigation. Applicants claiming such exemption shall submit a statement with their application to the Commission explaining the basis in detail for their finding (17.14 (a) of this chapter).

(2) Any antenna structure of 20 feet or less in height except one that would increase the height of another antenna structure (17.14 (b) of this chapter).

(c) All antennas (both receiving and transmitting) and supporting structures associated or used in conjunction with a Class C or D Citizens Radio Station operated from a fixed location must comply with at least one of the following:

(1) The antenna and its supporting structure does not exceed 20 feet in height above ground level; or

(2) The antenna and its supporting structure does not exceed by more than 20 feet the height of any natural formation, tree or man-made structure on which it is mounted; or

NOTE: A man-made structure is any construction other than a tower, mast, or pole.

(3) The antenna is mounted on the transmitting antenna structure of another authorized radio station and exceeds neither 60 feet above ground level nor the height of the antenna supporting structure of the other station; or

(4) The antenna is mounted on and does not exceed the height of the antenna structure otherwise used solely for receiving purposes, which structure itself complies with subparagraph (1) or (2) of this paragraph.

(5) The antenna is omnidirectional and the highest point of the antenna and its supporting structure does not exceed 60 feet above ground level and the highest point also does not exceed one foot in height above the established airport elevation for each 100 feet of horizontal distance from the nearest point of the nearest airport runway.

NOTE: A work sheet will be made available upon request to assist in determining the maximum permissible height of an antenna structure.

(d) Class C stations operated on frequencies in the 72—76 MHz band shall employ a transmitting antenna which complies with all of the following:

(1) The gain of the antenna shall not exceed that of a half-wave dipole;

(2) The antenna shall be immediately attached to, and an integral part of, the transmitter; and

(3) Only vertical polarization shall be used.

(e) Further details as to whether an aeronautical study and/or obstruction marking and lighting may be required, and specifications for obstruction marking and lighting when required, may be obtained from Part 17 of this chapter, "Construction, Marking, and Lighting of Antenna Structures."

(f) Subpart I of Part 1 of this chapter contains procedures implementing the National Environmental Policy Act of 1969. Applications for authorization of the construction of certain classes of communications facilities defined as "major actions" in 1.305 thereof, are required to be accompanied by specified statements. Generally these classes are:

(1) Antenna towers or supporting structures which exceed 300 feet in height and are not located in areas devoted to heavy industry or to agriculture.

(2) Communications facilities to be located in the following areas:

(i) Facilities which are to be located in an officially designated wilderness area or in an area whose designation as a wilderness is pending consideration;

(ii) Facilities which are to be located in an officially designated wildlife preserve or in an area whose designation as a wildlife preserve is pending consideration;

(iii) Facilities which will affect districts, sites, buildings, structures or objects, significant in American history, architecture, archaeology or culture, which are listed in the National Register of Historic Places or are eligible for listing (see 36 CFR 800.2 (d) and (f) and 800.10); and

(iv) Facilities to be located in areas which are recognized either nationally or locally for their special scenic or recreational value.

(3) Facilities whose construction will involve extensive change in surface features (e.g. wetland fill, deforestation or water diversion).

NOTE: The provisions of this paragraph do not include the mounting of FM, television or other antennas comparable thereto in size on an existing building or antenna tower. The use of existing routes, buildings and towers is an environmentally desirable alternative to the construction of new routes or towers and is encouraged.

If the required statements do not accompany the application, the pertinent facts may be brought to the attention of the Commission by any interested person during the course of the license term and considered de novo by the Commission.

[95.37 (c) & (c) (3) amended eff. 9-15-75; VI (75)-2]

SUBPART C—TECHNICAL REGULATIONS

95.41 Frequencies Available.

(a) Frequencies available for assignment to Class A stations:

(1) The following frequencies or frequency pairs are available primarily for assignment to base and mobile stations. They may also be assigned to fixed stations as follows:

(i) Fixed stations which are used to control base stations of a system may be assigned the frequency assigned to the mobile units associated with the base station. Such fixed stations shall comply with the following requirements if they are located within 75 miles of the center of urbanized areas of 200,000 or more population.

(a) If the station is used to control one or more base stations located within 45 degrees of azimuth, a directional antenna having a front-to-back ratio of at least 15 dB shall be used at the fixed station. For other situations where such a directional antenna cannot be used, a cardioid, bidirectional or omnidirectional antenna may be employed. Consistent with reasonable design, the antenna used must, in each case, produce a radiation pattern that provides only the coverage necessary to permit satisfactory control of each base station and limit radiation in other directions to the extent feasible.

(b) The strength of the signal of a fixed station controlling a single base station may not exceed the signal strength produced at the antenna terminal of the base receiver by a unit of the associated mobile station, by more than 6 dB. When the station controls more than one base station, the 6 dB control-to-mobile signal difference need be verified at only one of the base station sites. The measurement of the signal strength of the mobile unit must be made when such unit is transmitting from the control station location or, if that is not practical, from a location within one-fourth mile of the control station site.

(c) Each application for a control station to be authorized under the provisions of this paragraph shall be accompanied by a statement certifying that the output power of the proposed station transmitter will be adjusted to comply with the foregoing signal level limitation. Records of the measurements used to determine the signal ratio shall be kept with the station records and shall be made available for inspection by Commission personnel upon request.

(d) Urbanized areas of 200,000 or more population are defined in the U.S. Census of Population, 1960, Vol. 1, table 23, page 50. The centers of urbanized areas are determined from the Appendix, page 226 of the U.S. Commerce publication "Air Line Distance Between Cities in the United States."

(ii) Fixed stations, other than those used to control base stations, which are located 75 or more miles from the center of an urbanized area of 200,000 or more population. The centers of urbanized areas of 200,000 or more population are listed on page 226 of the Appendix to the U.S. Department of Commerce publication "Air Lines Distance Between Cities in the United States." When the fixed station is located 100 miles or less from the center of such an urbanized area, the power output may not exceed 15 watts. All fixed systems are limited to a maximum of two frequencies and must employ directional antennas with a front-to-back ratio of at least 15 dB. For two-frequency systems, separation between transmit-receive frequencies is 5 MHz.

Base and Mobile (MHz)	Mobile Only (MHz)
462.550	467.550
462.575	467.575
462.600	467.600
462.625	467.625
462.650	467.650
462.675	467.675
462.700	467.700
462.725	467.725

(2) Conditions governing the operation of stations authorized prior to March 18, 1968:

(i) All base and mobile stations authorized to operate on frequencies other than those listed in subparagraph (1) of this paragraph may continue to operate on those frequencies only until January 1, 1970.

(ii) Fixed stations located 100 or more miles from the center of any urbanized area of 200,000 or more population authorized to operate on frequencies other than those listed in subparagraph (1) of this paragraph will not have to change frequencies provided no interference is caused to the operation of stations in the land mobile service.

(iii) Fixed stations, other than those used to control base stations, located less than 100 miles (75 miles if the transmitter power output does not exceed 15 watts) from the center of any urbanized area of 200,000 or more population must discontinue operation by November 1, 1971. However, any operation after January 1, 1970, must be on frequencies listed in subparagraph (1) of this paragraph.

(iv) Fixed stations, located less than 100 miles from the center of any urbanized area of 200,000 or more population, which are used to control base stations and are authorized to operate on frequencies other than those listed in subparagraph (1) of this paragraph may continue to operate on those frequencies only until January 1, 1970.

(v) All fixed stations must comply with the applicable technical requirements of subparagraph (1) relating to antennas and radiated signal strength of this paragraph by November 1, 1971.

(vi) Notwithstanding the provisions of subdivisions (i) through (v) of this subparagraph, all stations authorized to operate on frequencies between 465.000 and 465.500 MHz and located within 75 miles of the center of the 20 largest urbanized areas of the United States, may continue to operate on these frequencies only until January 1, 1969. An extension to continue operation on such frequencies until January 1, 1970, may be granted to such station licensees on a case by case basis if the Commission finds that continued operation would not be inconsistent with planned usage of the particular frequency for police purposes. The 20 largest urbanized areas can be found in the U.S. Census of Population, 1960, vol. 1, table 23, page 50. The centers of urbanized areas are determined from the appendix, page 226, of the U.S. Commerce publication, "Air LIne Distance Between Cities in the United States."

(b) [Reserved]

(c) Class C mobile stations may employ only amplitude tone modulation or on-off keying of the unmodulated carrier, on a shared basis with other stations in the Citizens Radio Service on the frequencies and under the conditions specified in the following tables:

(1) For the control of remote objects or devices by radio, or for the remote actuation of devices which are used solely as a means of attracting attention and subject to no protection from interference due to the operation of industrial, scientific, or medical devices within the 26.96—27.28 MHz band, the following frequencies are available:

(MHz)	(MHz)	(MHz)
26.995	27.095	27.195
27.045	27.145	¹27.255

¹The frequency 27.255 MHz also is shared with stations in other services.

(2) Subject to the conditions that interference will not be caused to the remote control of industrial equipment operating on the same or adjacent frequencies and to the reception of television transmissions on Channels 4 or 5; and that no protection will be afforded from interference due to the operation of fixed and mobile stations in other services assigned to the same or adjacent frequencies in the band, the following frequencies are available solely for the radio remote control of models used for hobby pby purposes:

(i) For the radio remote control of any model used for hobby purposes:

MHz	MHz	MHz
72.16	72.32	72.89

(ii) For the radio remote control of aircraft models only:

MHz	MHz	MHz
72.08	72.24	72.40
75.64		

(d) The frequencies listed in the following paragraphs are available for use by Class D stations and are subject to no protection from interference resulting from the operation of industrial, scientific, or medical devices in the 26.96 MHz to 27.28 MHz band.

(1) The following frequencies may be used for communications between Class D stations:

MHz	MHz
26.965	27.115
26.975	27.125
26.985	27.135
27.005	27.155
27.015	27.165
27.025	27.175
27.035	27.185
27.055	27.205
27.075	27.214
27.085	27.215
27.105	27.225
	27.255

(2) Effective January 1, 1977, the following frequencies may be used for communications between Class D stations:

MHz	MHz
26.965	27.225
26.975	27.235
26.985	27.245
27.005	27.255
27.015	27.265
27.025	27.275
27.035	27.285
27.055	27.295
27.075	27.305
27.085	27.315
27.105	27.325
27.115	27.335
27.125	27.345
27.135	27.355
27.155	27.365
27.165	27.375
27.175	27.385
27.185	27.395
27.205	27.405
27.215	

(3) The frequency 27.065 MHz shall be used solely for:

(i) Emergency communications involving the immediate safety of life of individuals or the immediate protection of property, or

(ii) Communications necessary to render assistance to a motorist.

NOTE—A licensee, before using 27.065 MHz must make a determination that his communication is either or both (a) an emergency communication or (b) is necessary to render assistance to a motorist. To be an emergency communication, the message must have some direct relation the the immediate safety of life or immediate protection of property. If no immediate action is required, it is not an emergency. What may not be an emergency under one set of circumstances may be an emergency under different circumstances. There are many worthwhile public service communications that do not qualify as emergency communications. In the case of motorist assistance, the message must be necessary to assist a particular motorist and not, except in a valid emergency, motorists in general. If the communications are to be lengthy, the exchange should be shifted to another frequency, if feasible, after contact is established. No nonemergency or nonmotorist assistance communications are permitted on 27.065 MHz even for the limited purpose of calling a licensee monitoring a frequency to ask him to switch to another frequency. Although 27.065 MHz may be used for marine emergencies, it should not be considered a substitute for the authorized marine distress system. The Coast Guard has stated it will not "participate directly in the Citizens Radio Service by fitting with and/or providing a watch on any Citizens Band Channel. (Coast Guard Commandant Instructions 2302.6)"

The following are examples of permitted and prohibited types of communications. They are guidelines and are not intended to be all inclusive.

Permitted	Example message
Yes.........	A tornado is sighted six miles north of town.
No	This is observation post number 10. No tornados sighted.
Yes.........	I am out of gas on Interstate 95.
No	I am out of gas in my driveway.
Yes.........	There is a four-car collision at Exit 10 on the Beltway.
No	Traffic is moving smoothly on the Beltway.
Yes.........	Base to Unit 1, the Weather Bureau has just issued a thunderstorm warning. Bring the sailboat into port.

No	Attention all motorists. The Weather Bureau advises that the snow tomorrow will accumulate 4 to 6 inches.
Yes	There is a fire in the building on the corner of 6th and Main Streets.
No	This is Halloween patrol unit number 3. Everything is quiet here.

The following priorities should be observed in the use of 27.065 MHz:

1. Communications relating to an existing situation dangerous to life or property, i.e., fire, automobile accident.
2. Communications relating to a potentially hazardous situation, i.e., car stalled in a dangerous place, lost child, boat out of gas.
3. Road assistance to a disabled vehicle on the highway or street.
4. Road and street directions.

(e) Upon specific request accompanying application for renewal of station authorization, a Class A station in this service, which was authorized to operate on a frequency in the 460—461 MHz band until March 31, 1967, may be assigned that frequency for continued use until not later than March 31, 1968, subject to all other provisions of this part.

[95.41 (d) & (d) (1) amended, (d) (2) deleted, (d) (3) redesig. (d) (2), and new (d) (3) added eff. 9-15-75; VI (75)-2]

95.42 Special provisions.

Effective September 10, 1976 station authorizations for the use of frequencies between 26.96 MHz and 27.41 MHz will be issued only to applicants in the Citizens Radio Service. Any license in a radio service other than the Citizens Radio Service authorizing the use of frequencies between 26.96 MHz and 27.41 MHz shall remain valid until December 31, 1979.

95.43 Transmitter power.

(a) Transmitter power is the power at the transmitter output terminals and delivered to the antenna, antenna transmission line, or any other impedance matched, radio frequency load.

(1) For single sideband transmitters and other transmitters employing a reduced carrier, a suppressed carrier or a controlled carrier, used at Class D stations, transmitter power is the peak envelope power.

(2) For all transmitters other than those covered by paragraph (a) (1) of this section, the transmitter power is the carrier power.

(b) The transmitter power of a station shall not exceed the following values under any condition of modulation or other circumstances.

Class of station:	Transmitter power in watts
A	50
C—27.255 MHz	25
C—26.995—27.195 MHz	4
C—72—76 MHz	0.75
D—Carrier (where applicable)	4
D—Peak envelope power (where applicable)	12

95.44 External radio frequency power amplifiers prohibited.

No external radio frequency power amplifier shall be used or attached, by connection, coupling attachment or in any other way at any Class D station.

NOTE: An external radio frequency power amplifier at a Class D station will be presumed to have been used where it is in the operator's possession or on his premises and there is extrinsic evidence of any operation of such Class D station in excess of power limitations provided under this rule part unless the operator of such equipment holds a station license in another radio service under which license the use of the said amplifier at its maximum rated output power is permitted.

95.45 Frequency tolerance.

(a) Except as provided in paragraphs (b) and (c) of this section, the carrier frequency of a transmitter in this service shall be maintained within the following percentage of the authorized frequency:

Frequency tolerance

Class of station	Fixed and base	Mobile
A	0.0025	0.0005
C		.005
D		.005

(b) Transmitters used at Class C stations operating on authorized frequencies between 26.99 and 27.26 MHz with 2.5 watts or less mean output power, which are used solely for the control of remote objects or devices by radio (other than devices used solely as a means of attracting attention), are permitted a frequency tolerance of 0.01 percent.

(c) Class A stations operated at a fixed location used to control base stations, through use of a mobile only frequency, may operate with a frequency tolerance of 0.0005 percent.

95.47 Types of emission.

(a) Except as provided in paragraph (e) of this section, Class A stations in this service will normally be authorized to transmit radiotelephony only. However, the use of tone signals or signalling devices solely to actuate receiver circuits, such as tone operated squelch or selective calling circuits, the primary function of which is to establish or establish and maintain voice communications, is permitted. The use of tone signals solely to attract attention is prohibited.

(b) [Reserved]

(c) Class C stations in this service are authorized to use amplitude tone modulation or on-off unmodulated carrier only, for the control of remote objects or devices by radio or for the remote actuation of devices which are used solely as a means of attracting attention. The transmission of any form of telegraphy, telephony or record communications by a Class C station is prohibited. Telemetering, except for the transmission of simple, short duration signals indicating the presence or absence of a condition or the occurrence of an event, is also prohibited.

(d) Transmitters used at Class D stations in this service are authorized to use amplitude voice modulation, either single or double sideband. Tone signals or signalling devices may be used only to actuate receiver circuits, such as tone operated squelch or selective calling circuits, the primary function of which is to establish or maintain voice communications. The use of any signals solely to attract attention or for the control of remote objects or devices is prohibited.

(e) Other types of emission not described in paragraph (a) of this section may be authorized for Class A citizens radio stations upon a showing of need therefor. An application requesting such authorization shall fully describe the emission desired, shall indicate the bandwidth required for satisfactory communication, and shall state the purpose for which such emission is required. For information regarding the classification of emissions and the calculation of bandwidth, reference should be made to Part 2 of this chapter.

95.49 Emission limitations.

(a) Each authorization issued to a Class A citizens radio station will show, as a prefix to the classification of the authorized emission, a figure specifying the maximum bandwidth to be occupied by the emission.

(b) [Reserved]

(c) The authorized bandwidth of the emission of any transmitter employing amplitude modulation shall be 8 kHz for double sideband and 4 kHz for single sideband. The authorized bandwidth of the emission of any transmitter employing frequency or phase modulation (Class F2 or F3) shall be 20 kHz. The use of F2 and F3 emissions in the frequency band 26.96 MHz—27.41 is not authorized.

(d) The mean power of emissions shall be attenuated below the mean power of the transmitter in accordance with the following schedule:

(1) When using emissions other than single sideband:

(i) On any frequency removed from the center of the authorized bandwidth by more than 50 percent up to and including 100 percent of the authorized bandwidth: at least 25 decibels;

(ii) On any frequency removed from the center of the authorized bandwidth by more than 100 percent up to and including 250 percent of the authorized bandwidth; At least 35 decibels;

(2) When using single sideband emissions:

(i) On any frequency removed from the center of the authorized bandwidth by more than 50 percent up to and including 150 percent of the authorized bandwidth: At least 25 decibels:

(ii) On any frequency removed from the center of the authorized bandwidth by more than 150 percent up to and including 250 percent of the authorized bandwidth: At least 35 decibels;

(3) On any frequency removed from the center of the authorized bandwidth by more than 250 percent of the authorized bandwidth: at least $43 + 10 \log_{10} 10$ (mean power in watts) decibels, for Class D transmitters type accepted before September 10, 1976 and all Class A transmitters.

(4) On any frequency removed from the center of the authorized bandwidth by more than 250 percent of the authorized bandwidth up to a frequency of twice the fundamental frequency; at least $53 + 10 \log_{10}$ (mean power in watts) decibels, for Class D transmitters type accepted after September 10, 1976.

(5) On any frequency twice or greater than twice the fundamental frequency: at least 60 decibels (mean power in watts) for Class D transmitters type accepted after September 10, 1976.

NOTE—The requirements of paragraph (d) must be met both with and without connection of all attachments acceptable for use with such transmitters. External speakers, microphones, power cords, and antennas are among the devices included in this requirement. Additionally, if it is shown that a licensee causes interference to television reception because of insufficient harmonic attenuation, he may be required to insert a low pass filter between the transmitter RF output terminal and the antenna feedline.

(e) When an unauthorized emission results in harmful interference, the Commission may, in its discretion, require appropriate technical changes in equipment to alleviate the interference.

95.51 Modulation requirements.

(a) When double sideband, amplitude modulation is used for telephony, the modulation percentage shall be sufficient to provide efficient communication and shall not exceed 100 percent.

(b) Each transmitter for use in Class D stations, other than single sideband, suppressed carrier, or controlled carrier, for which type acceptance is requested after May 24, 1974, having more than 2.5 watts maximum output power shall be equipped with a device which automatically prevents modulation in excess of 100 percent on positive and negative peaks.

(c) The maximum audio frequency required for satisfactory radiotelephone intelligibility for use in this service is considered to be 3000 Hz.

(d) Transmitters for use at Class A stations shall be provided with a device which automatically will prevent greater than normal audio level from causing modulation in excess of that specified in this subpart; *Provided, however,* That the requirements of this paragraph shall not apply to transmitters authorized at mobile stations and having an output power of 2.5 watts or less.

(e) Each transmitter of a Class A station which is equipped with a modulation limiter in accordance with the provisions of paragraph (d) of this section shall also be equipped with an audio low-pass filter. This audio low-pass filter shall be installed between the modulation limiter and the modulated stage and, at audio frequencies between 3 kHz and 20 kHz, shall have an attenuation greater than the attenuation at 1 kHz by at least:

$$60 \log_{10} (f/3) \text{ decibels}$$

where "f" is the audio frequency in kHz. At audio frequencies above 20 kHz, the attenuation shall be at least 50 decibels greater than the attenuation at 1 kHz.

(f) Simultaneous amplitude modulation and frequency or phase modulation of a transmitter is not authorized.

(g) The maximum frequency deviation of frequency modulated transmitters used at Class A stations shall not exceed $+5$ kHz.

95.53 Compliance with technical requirements.

(a) **Upon receipt of notification from the Commission of a deviation from the technical requirements of the rules in this part, the radiations of the transmitter involved shall be suspended immediately, except for necessary tests and adjustments, and shall not be resumed until such deviation has been corrected.**

(b) **When any citizens radio station licensee receives a notice of violation indicating that the station has been operated contrary to any of the provisions contained in Subpart C of this part, or where it otherwise appears that operation of a station in this service may not be in accordance with applicable technical standards, the Commission may require the licensee to conduct such tests as may be necessary to determine whether the equipment is capable of meeting these standards and to make such adjustments as may be necessary to assure compliance therewith. A licensee who is notified that he is required to conduct such tests and/or make adjustments must, within the time limit specified in the notice, report to the Commission the results thereof.**

(c) All tests and adjustments which may be required in accordance with paragraph (b) of this section shall be made by, or under the immediate supervision of, a person holding a first- or second-class commercial operator license, either radiotelephone or radio telegraph as may be appropriate for the type of emission employed. In each case, the report which is submitted to the Commission shall be signed by the licensed commercial operator. Such report shall describe the results of the tests and adjustments, the test equipment and procedures used, and shall state the type, class, and serial number of the operator's license. A copy of this report shall also be kept with the station records.

95.55 Acceptability of transmitters for licensing.

Transmitters type approved or type accepted for use under this part are included in the Commission's Radio Equipment List. Copies of this list are available for public reference at the Commission's Washington, D.C., offices and field offices. The requirements for transmitters which may be operated under a license in this service are set forth in the following paragraphs.

(a) Class A stations: All transmitters shall be type accepted.

(b) Class C stations:

(1) Transmitters operated in the band 72—76 MHz shall be type accepted.

(2) All transmitters operated in the band 26.99—27.26 MHz shall be type approved, type accepted or crystal controlled.

(c) Class D stations:

(1) All transmitters first licensed, or marketed as specified in 2.805 of this chapter, prior to November 22, 1974 shall be type accepted or crystal controlled.

(2) All transmitters first licensed, or marketed as specified in 2.803 of this chapter, on or after November 22, 1974, shall be type accepted.

(3) Effective November 23, 1978, all transmitters shall be type accepted.

(4) Prior to January 1, 1977 transmitters which are equipped to operate on any frequency not included in 95.41 (d) (1) may not be installed at, or used by, any Class D station unless there is a station license posted at the transmitter location, or a transmitter identification card (FCC Form 452—C) attached to the transmitter, which indicates that operation of the transmitter on such frequency has been authorized by the Commission.

(5) Effective January 1, 1977 transmitters which are equipped to operate on any frequency not included in 95.41 may not be installed at or used by any Class D station unless there is a station license posted at the transmitter location, or a transmitter identification card (FCC Form 452—C) attached to the transmitter, which indicates that operation of the transmitter on such frequency has been authorized by the Commission.

NOTE—A "transmitter" is defined to include any radio frequency (RF) power amplifier.

(6) No Class D transmitter type accepted prior to September 10, 1976 shall be manufactured on or after August 1, 1977.

(7) No Class D transmitter type accepted prior to September 10, 1976 shall be marketed on or after January 1, 1978.

(d) With the exception of equipment type approved for use at a Class C station, all transmitting equipment authorized in this service shall be crystal controlled.

(e) No controls, switches or other functions which can cause operation in violation of the technical regulations of this part shall be accessible from the operating panel or exterior to the cabinet enclosing a transmitter authorized in this service.

95.57 Procedure for type acceptance of equipment.

(a) Any manufacturer of a transmitter built for use in this service, except noncrystal controlled transmitters for use at Class C stations, may request type acceptance for such transmitter in accordance with the type acceptance requirements of this part, following the type acceptance procedure set forth in Part 2 of this chapter.

(b) Type acceptance for an individual transmitter may also be requested by an applicant for a station authorization by following the type acceptance procedures set forth in Part 2 of this chapter. Such transmitters, if accepted, will not normally be included on the Commission's "Radio Equipment List" but will be individually enumerated on the station authorization.

(c) Additional rules with respect to type acceptance are set forth in Part 2 of this chapter. These rules include information with respect to withdrawal of type acceptance, modification of type-accepted equipment, and limitations on the findings upon which type acceptance is based.

(d) Transmitters equipped with a frequency or frequencies not listed in 95.41(d) (1) will not be type accepted for use at Class D stations unless the transmitter is also type accepted for use in the service in which the frequency is authorized, if type acceptance in that service is required.

95.58 Additional requirements for type acceptance.

(a) All transmitters shall be crystal controlled.

(b) Except for transmitters type accepted for use at Class A stations, transmitters shall not include any provisions for increasing power to levels in excess of the pertinent limits specified in Section 95.43.

(c) In addition to all other applicable technical requirements set forth in this part, transmitters for which type acceptance is requested after May 24, 1974, for use at Class D stations shall comply with the following:

(1) Single sideband transmitters and other transmitters employing reduced, suppressed or controlled carrier shall include a means for automatically preventing the transmitter power from exceeding either the maximum permissible peak envelope power or the rated peak envelope power of the transmitter, whichever is lower.

(2) Multi-frequency transmitters shall be capable of operation only on those frequencies authorized by 95.41.

(3) All transmitter frequency determining circuitry (including crystals), other than the frequency selection mechanism, employed in Class D station equipment shall be internal to the equipment and shall not be accessible from the exterior of the equipment cabinet or operating panel. Add-on devices, whether internal or external to the equipment, the function of which is to extend the frequency coverage capability of a Class D unit beyond its original frequency coverage capability, shall not be sold, manufactured, or attached to any transmitter capable of operation on Class D Citizens Radio Service frequencies.

(4) Single sideband transmitters shall be capable of transmitting on the upper sideband. Capability for transmission also on the lower sideband is permissible.

(5) The total dissipation ratings, established by the manufacturer of the electron tubes or semiconductors which supply radio frequency power to the antenna terminals of the transmitter, shall not exceed 10 watts. For electron tubes, the rating shall be the Intermittent Commercial and Amateur Service (ICAS) plate dissipation value if established. For semiconductors, the rating shall be the collector or device dissipation value, whichever is greater, which may be temperature de-rated to not more than 50°C.

(d) Only the following external transmitter controls, connections or devices will normally be permitted in transmitters for which type acceptance is requested after May 24, 1974, for use at Class D stations. Approval of additional controls, connections or devices may be given after consideration of the function to be performed by such additions.

(1) Primary power connection. (Circuitry or devices such as rectifiers, transformers, or inverters which provide the nominal rated transmitter primary supply voltage may be used without voiding the transmitter type acceptance.)

(2) Microphone connection.

(3) Radio frequency output power connection.

(4) Audio frequency power amplifier output connector and selector switch.

(5) On-off switch for primary power to transmitter. May be combined with receiver controls such as the receiver controls such as the receiver on-off switch and volume control.

(6) Upper-lower sideband selector; for single sideband transmitters only.

(7) Selector for choice of carrier level; for single sideband transmitters only. May be combined with sideband selector.

(8) Transmitting frequency selector switch.

(9) Transmit-recieve switch.

(10) Meter(s) and selector switch for monitoring transmitter performance.

(11) Pilot lamp or meter to indicate the presence of radio frequency output power or that transmitter control circuits are activated to transmit.

(e) An instruction book for the user shall be furnished with each transmitter sold and one copy (a draft or preliminary copy is acceptable providing a final copy is furnished when completed) shall be forwarded to the Commission with each request for type acceptance or type approval. The book shall contain all information necessary for the proper installation and operation of the transmitter including:

(1) Instructions concerning all controls, adjustments and switches which may be operated or adjusted without causing violation of techincal regulations of this part;

(2) Warnings concerning any adjustment which, according to the rules of this part, may be made only by, or under the immediate supervision of, a person holding a commercial first or second class radio operator license;

(3) Warnings concerning the replacement or substitution of crystals, tubes or other components which could cause violation of the technical regulations of this part and of the type acceptance or type approval requirements of Part 2 of this chapter.

(4) Warnings concerning licensing requirements and details concerning the application procedures for licensing.

(f) A Class D Citizens Radio Service application form (FCC Form 505), a Temporary Permit, Class D Citizens Radio Station (FCC Form 555—B), and a copy of Part 95 of the Commission's Rules and Regulations, each to be current at the time of packing of the transmitter, shall be furnished with each transmitter sold after January 1, 1977.

(g) The serial number of each new Class D unit sold after January 1, 1977 shall be engraved on the unit's chassis.

95.59 Submission of noncrystal controlled Class C station transmitters for type approval.

Type approval of noncrystal controlled transmitters for use at Class C stations in this service may be requested in accordance with the procedure specified in Part 2 of this chapter.

95.61 Type approval of receiver-transmitter combinations.

Type approval will not be issued for transmitting equipment for operation under this part when such equipment is enclosed in the same cabinet, is constructed on the same chassis in whole or in part, or is identified with a common type or model number with a radio receiver; unless such receiver has been cerrificated to the Commission as complying with the requirements of Part 15 of this chapter.

95.63 Minimum equipment specifications.

Transmitters submitted for type approval in this service shall be capable of meeting the technical specifications contained in this part, and in addition, shall comply with the following:

(a) Any basic instructions concerning the proper adjustment, use, or operation of the equipment that may be necessary shall be attached to the equipment in a suitable manner and in such positions as to be easily read by the operator.

(b) A durable nameplate shall be mounted on each transmitter showing the name of the manufacturer, the type or model designation, and providing suitable space for permanently displaying the transmitter serial number, FCC type approval number, and the class of station for which approved.

(c) The transmitter shall be designed, constructed, and adjusted by the manufacturer to operate on a frequency or frequencies available to the class of station for which type approval is sought. In designing the equipment, every reasonable precaution shall be taken to protect the user from high voltage shock and radio frequency burns. Connections to batteries (if used) shall be made in such a manner as to permit replacement by the user without causing improper operation of the transmitter. Generally accepted modern engineering principles shall be utilized in the generation of radio frequency currents so as to guard against unnecessary interference to other services. In cases of harmful interference arising from the design, construction, or operation of the equipment, the Commission may require appropriate technical changes in equipment to alleviate interference.

(d) Controls which may effect changes in the carrier frequency of the transmitter shall not be accessible from the exterior of any unit unless such accessibility is specifically approved by the Commission.

95.65 Test procedure.

Type approval tests to determine whether radio equipment meets the technical specifications contained in this part will be conducted under the following conditions:

(a) Gradual ambient temperature variations from 0° to 125°F.

(b) Relative ambient humidity from 20 to 95 percent. This test will normally consist of subjecting the equipment for at least three consecutive periods of 24 hours each, to a relative ambient humidity of 20, 60 and 95 percent, respectively, at a temperature of approximately 80°F.

(c) Movement of transmitter or objects in the immediate vicinity thereof.

(d) Power supply voltage variations normally to be encountered under actual operating conditions.

(e) Additional tests as may be prescribed, if considered necessary or desirable.

95.67 Certificate of type approval.

A certificate or notice of type approval, when issued to the manufacturer of equipment intended to be used or operated in the Citizens Radio Service, constitutes a recognition that on the basis of the test made, the particular type of equipment appears to have the capability of functioning in accordance with the technical specifications and regulations contained in this part: *Provided,* That all such additional equipment of the same type is properly constructed, maintained, and operated: *And provided further,* That no change whatsoever is made in the design or construction of such equipment except upon specific approval by the Commission.

SUBPART D—STATION OPERATING REQUIREMENTS

95.81 Permissible Communications.

Stations licensed in the Citizens Radio Service are authorized to transmit the following types of communications:

(a) Communications to facilitate the personal or business activities of the licensee.

(b) Communication relating to:

(1) The immediate safety of life or the immediate protection of property in accordance with 95.85.

(2) The rendering of assistance to a motorist, mariner or other traveler.

(3) Civil defense activities in accordance with 95.121.

(4) Other activities only as specifically authorized pursuant to 95.87.

(c) Communications with stations authorized in other radio services except as prohibited in 95.83(a) (3).

95.81 added eff. 9-15-75; VI (75)-2]

95.83 Prohibited communications.

(a) A citizens radio station shall not be used:

(1) For any purpose, or in connection with any activity, which is contrary to Federal, State, or local law.

(2) For the transmission of communications containing obscene, indecent, profane words, language, or meaning.

(3) To communicate with an Amateur Radio Service station, an unlicensed station, or foreign stations (other than as provided in Subpart E of this part) except for communications pursuant to 95.85(b) and 95.121.

(4) To convey program material for retransmission, live or delayed, on a broadcast facility. *NOTE:* A Class A or Class D station may be used in connection with administrative, engineering, or maintenance activities of a broadcasting station; a Class A or Class C station may be used for control functions by radio which do not involve the transmission of program material; and a Class A or Class D station may be used in the gathering of news items or preparation of programs: Provided, that the actual or recorded transmissions of the Citizens radio station are not broadcast at any time in whole or in part.

(5) To intentionally interfere with the communications of another station.

(6) For the direct transmission of any material to the public through a public address system or similar means.

(7) For the transmission of music, whistling, sound effects, or any material for amusement or entertainment purposes, or solely to attract attention.

(8) To transmit the word "MAYDAY" or other international distress signals, except when the station is located in a ship, aircraft, or other vehicle which is threatened by grave and imminent danger and requests immediate assistance.

(9) For advertising or soliciting the sale of any goods or services.

(10) For transmitting messages in other than plain language. Abbreviations including nationally or internationally recognized operating signals, may be used only if a list of all such abbreviations and their meaning is kept in the station records and made available to any Commission representative on demand.

(11) To carry on communications for hire, whether the remuneration or benefit received is direct or indirect.

[95.83(a) & headnote amended eff. 9-15-75; VI (75)-2]

(b) A Class D station may not be used to communicate with, or attempt to communicate with, any unit of the same or another station over a distance of more than 150 miles.

(c) A licensee of a Citizens radio station who is engaged in the business of selling Citizens radio transmitting equipment shall not allow a customer to operate under his station license. In addition, all communications by the licensee for the purpose of demonstrating such equipment shall consist only of brief messages addressed to other units of the same station.

95.85 Emergency and assistance to motorist use.

(a) All Citizens radio stations shall give priority to the emergency communications of other stations which involve the immediate safety of life of individuals or the immediate protection of property.

(b) Any station in this service may be utilized during an emergency involving the immediate safety of life of individuals or the immediate protection of property for the transmission of emergency communications. It may also be used to transmit communications necessary to render assistance to a motorist.

(1) When used for transmission of emergency communications certain provisions in this part concerning use of frequencies (95.41(d); prohibited uses (95.83(a) (3); operation by or on behalf of persons other than the licensee (95.87); and duration of transmissions (95.91 (a) and (b) shall not apply.

(2) When used for transmissions of communications necessary to render assistance to a traveler, the provisions of this Part concerning duration of transmission (95.91(b) shall not apply.

(3) The exemptions granted from certain rule provisions in subparagraphs (1) and (2) of this paragraph may be rescinded by the Commission at its discretion.

(c) If the emergency use under paragraph (b) of this section extends over a period of 12 hours or more, notice shall be sent to the Commission in Washington, D.C., as soon as it is evident that the emergency has or will exceed 12 hours. The notice should include the identity of the stations

participating, the nature of the emergency, and the use made of the stations. A single notice covering all participating stations may be submitted.
[95.85(b) (1) & (2) amended eff. 9-15-75; VI (75)-2]
95.87 Operation by, or on behalf of, persons other than the licensee.
(a) Transmitters authorized in this service must be under the control of the licensee at all times. A licensee shall not transfer, assign, or dispose of, in any manner, directly or indirectly, the operating authority under his station license, and shall be responsible for the proper operation of all units of the station.

(b) Citizens radio stations may be operated only by the following persons, except as provided in paragraph (c) of this section:
 (1) The licensee;
 (2) Members of the licensee's immediate family living in the same household;
 (3) The partners, if the licensee is a partnership, provided the communications relate to the business of the partnership;
 (4) The members, if the licensee is an unincorporated association, provided the communications relate to the business of the association;
 (5) Employees of the licensee only while acting within the scope of their employment;
 (6) Any person under the control or supervision of the licensee when the station is used solely for the control of remote objects or devices, other than devices used only as a means of attracting attention; and
 (7) Other persons, upon specific prior approval of the Commission shown on or attached to the station license, under the following circumstances:
  (i) LIcensee is a corporation and proposes to provide private radiocommunication facilities for the transmission of messages or signals by or on behalf of its parent corporation, another subsidiary of the parent corporation, or its own subsidiary. Any remuneration or compensation received by the licensee for the use of the radio communication facilities shall be governed by a contract entered into by the parties concerned and the total of the compensation shall not exceed the cost of providing the facilities. Records which show the cost of service and its nonprofit or cost-sharing basis shall be maintained by the licensee.
  (ii) Licensee proposes the shared or cooperative use of a Class A station with one or more other licensees in this service for the purpose of communicating on a regular basis with units of their respective Class A stations, or with units of other Class A stations if the communications transmitted are otherwise permissible. The use of these private radiocommunication facilities shall be conducted pursuant to a written contract which shall provide that contributions to capital and operating expense shall be made on a nonprofit, cost-sharing basis, the cost to be divided on an equitable basis among all parties to the agreement. Records which show the cost of service and its nonprofit, cost-sharing basis shall be maintained by the licensee. In any case, however, licensee must show a separate and independent need for the particular units proposed to be shared to fulfill his own communications requirements.
  (iii) Other cases where there is a need for other persons to operate a unit of licensee's radio station. Requests for authority may be made either at the time of the filing of the application for station license or thereafter by letter. In either case, the licensee must show the nature of the proposed use and that it relates to an activity of the licensee, how he proposes to maintain control over the transmitters at all times, and why it is not appropriate for such other person to obtain a station license in his own name. The authority, if granted, amy be specific with respect to the names of the persons who are permitted to operate, or may authorize operation by unnamed persons for specific purposes. This authority may be revoked by the Commission, in its descretion, at any time.

(c) An individual who was formerly a citizens radio station licensee shall not be permitted to operate any citizens radio station of the same class licensed to another person until such time as he again has been issued a valid radio station license of that class, when his license has been:
 (1) Revoked by the Commission.
 (2) Surrendered for cancellation after the institution of revocation proceedings by the Commission.
 (3) Surrendered for cancellation after a notice of apparent liability to forfeiture has been served by the Commission.

95.89 Telephone answering services.
(a) Notwithstanding the provisions of 95.87, a licensee may install a transmitting unit of his station on the premises of a telephone answering service. The same unit may not be operated under the authorization of more than one licensee. In all cases, the licensee must enter into a written agreement with the answering service. This agreement must be kept with the licensee's station records and must provide, as a minimum, that:
 (1) The licensee will have control over the operation of the radio unit at all times;
 (2) The licensee will have full and unrestricted access to the transmitter to enable him to carry out his responsibilities under his license;
 (3) Both parties understand that the licensee is fully responsible for the proper operation of the citizens radio station; and
 (4) The unit so furnished shall be used only for the transmission of communications to other units belonging to the licensee's station.

(b) A citizens radio station licensed to a telephone answering service shall not be used to relay messages or transmit signals to its customers.

95.91 Duration of transmissions.
(a) All communications or signals, regardless of their nature, shall be restricted to the minimum practicable transmission time. The radiation of energy shall be limited to transmissions modulated or keyed for actual permissible communications, tests, or control signals. Continuous or uninterrupted transmissions from a single station or between a number of communicating stations is prohibited, except for communications involving the immediate safety of life or property.

(b) All communications between Class D stations (interstation) shall be restricted to not longer than five (5) continuous minutes. At the conclusion of this 5 minute period, or the exchange of less than 5 minutes, the participating stations shall remain silent for at least one minute.

(c) All communication between units of the same Class D station (intrastation) shall be restricted to the minimum practicable transmission.

(d) The transmission of audible tone signals or a sequency of tone signals for the operation of the tone operated squelch or selective calling circuits in accordance with 95.47 shall not exceed a total of 15 seconds duration. Continuous transmission of a subaudible tone for this purpose is permitted. For the purposes of this section, any tone or combination of tones having no frequency above 150 hertz shall be considered subaudible.

(e) The transmission of permissible control signals shall be limited to the minimum practicable time necessary to accomplish the desired control or actuation of remote objects or devices. The continuous radiation of energy for periods exceeding 3 minutes duration for the purpose of transmission of control signals shall be limited to control functions requiring at least one or more changes during each minute of such transmission. However, while it is actually being used to control model aircraft in flight by means of interrupted tone modulation of its carrier, a citizens radio station may transmit a continuous carrier without being simultaneously modulated if the presence or absence of the carrier also performs a control function. An exception to the limitations contained in this paragraph may be authorized upon a satisfactory showing that a continuous control signal is required to perform a control function which is necessary to insure the safety of life or property.
[95.91lb) amended & (c) added, present par. (c) & (d) redesig. par. (d) & (e) eff. 9-15-75; VI (75)-2]

95.93 Tests and adjustments.
All tests or adjustments of citizens radio transmitting equipment involving an external connection to the radio frequency output circuit shall be made using a nonradiating dummy antenna. However, a brief test signal, either with or without modulation, as appropriate, may be transmitted when it is necessary to adjust a transmitter to an antenna for a new station installation or for an existing installation involving a change of antenna or change of transmitters, or when necessary for the detection, measurement, and suppression of harmonic or other spurious radiation. Test transmissions using a radiating antenna shall not exceed a total of 1 minute during any 5-minute period, shall not interfere with communications already in progress on the operating frequency, and shall be properly identified as required by 95.95, but may otherwise be unmodulated as appropriate.

95.95 Station identification.
(a) The call sign of a citizens radio station shall consist of three letters followed by four digits.

(b) Each transmission of the station call sign shall be made in the English language by each unit, shall be complete, and each letter and digit shall be separately and distinctly transmitted. Only standard phonetic alphabets, nationally or internationally recognized, may be used in lieu of pronunciation of letters for voice transmission of call signs. A unit designator or special identification may be used in addition to the station call sing but not as a substitute therefor.

(c) Except as provided in paragraph (d) of this section, all transmission from each unit of a citizens radio station shall be identified by the transmission of its assigned call sign at the beginning and end of each transmission or series of transmissions, but at least at intervals not to exceed ten (10) minutes.

(d) Unless specifically required by the station authorization, the transmissions of a citizens radio station need not be identified when the station (1) is a Class A station which automatically retransmits the information received by radio from another station which is properly identified or (2) is not being used for telephony emission.

(e) In lieu of complying with the requirements of paragraph (c) of this section, Class A base stations, fixed stations, and mobile units when communicating with base stations may identify as follows:
 (1) Base stations and fixed stations of a Class A radio system shall transmit their call signs at the end of each transmission or exchange of transmissions, or once each 15-minute period of a continuous exchange of communications.
 (2) A mobile unit of a Class A station communicating with a base station of a Class A radio system on the same frequency shall transmit once during each exchange of transmissions any unit identifier which is on file in the station records of such base station.
 (3) A mobile unit of Class A stations communicating with a base station of a Class A radio system on a different frequency shall transmit its call sign at the end of each transmission or exchange of transmissions, or once each 15-minute period of a continuous exchange of communications.

[95.95(c) amended eff. 9-15-75; VI (75)-2]
95.97 Operator license requirements.

(a) No operator license is required for the operation of a citizens radio station except that stations manually transmitting Morse Code shall be operated by the holders of a third or higher class radiotelegraph operator license.

(b) Except as provided in paragraph (c) of this section, all transmitter adjustments or tests while radiating energy during or coincident with the construction, installation, servicing, or maintenance of a radio station in this service, which may affect the proper operation of such stations, shall be made by or under the immediate supervision and responsibility of a person holding a first- or second-class commercial radio operator license, either radiotelephone or radio telegraph, as may be appropriate for the type of emission employed, and such person shall be responsible for the proper functioning of the station equipment at the conclusion of such adjustments or tests. Further, in any case where a transmitter adjustment which may affect the proper operation of the transmitter has been made while not radiating energy by a person not the holder of the required commercial radio operator license or not under the supervision of such licensed operator, other than the factory assembling or repair of equipment, the transmitter shall be checked for compliance with the technical requirements of the rules by a commercial radio operator of the proper grade before it is placed on the air.

(c) Except as provided in 95.53 and in paragraph (d) of this section, no commercial radio operator license is required to be held by the person performing transmitter adjustments or test during or coincident with the construction, installation, servicing, or maintenance of Class C transmitters, or Class D transmitters used at stations authorized prior to May 24, 1974: *Provided*, That there is compliance with all of the following conditions:

(1) The transmitting equipment shall be crystal controlled with a crystal capable of maintaining the station frequency within the prescribed tolerance;

(2) The transmitting equipment either shall have been factory assembled or shall have been provided in kit form by a manufacturer who provided all components together with full and detailed instructions for their assembly by nonfactory personnel;

(3) The frequency determining elements of the transmitter, including the crystal(s) and all other components of the crystal oscillator circuit, shall have been preassembled by the manufacturer, pretuned to a specific available frequency, and sealed by the manufacturer so that replacement of any component or any adjustment which might cause off-frequency operation cannot be made without breaking such seal and thereby voiding the certification of the manufacturer required by this paragraph;

(4) The transmitting equipment shall have been so designed that none of the transmitter adjustments or tests normally performed during or coincident with the installation, servicing, or maintenance of the station, or during the normal rendition of the service of the station, or during the final assembly of kits or partially preassembled units, may reasonably be expected to result in off-frequency operation, excessive input power, overmodulation, or excessive harmonics or other spurious emissions; and

(5) The manufacturer of the transmitting equipment or of the kit from which the transmitting equipment is assembled shall have certified in writing to the purchaser of the equipment (and to the Commission upon request) that the equipment has been designed, manufactured, and furnished in accordance with the specifications contained in the foregoing subparagraphs of this paragraph. The manufacturer's certification concerning design and construction features of Class C or Class D station transmitting equipment, as required if the provisions of this paragraph are invoked, may be specific as to a particular unit of transmitting equipment or general as to a group or model of such equipment, and may be in any form adequate to assure the purchase of the equipment or the Commission that the conditions described in this paragraph have been fulfilled.

(d) Any tests and adjustments necessary to correct any deviation of a transmitter of any Class of station in this service from the technical requirements of the rules in this part shall be made by, or under the immediate supervision of, a person holding a first- or second-class commercial operator license, either radiotelephone or radiotelegraph, as may be appropriate for the type of emission employed.

95.101 Posting station license and transmitter identification cards or plates.

(a The current authorization, or a clearly legible photocopy thereof, for each station (including units of a Class C or Class D station) operated at a fixed location shall be posted at a conspicuous place at the principal fixed location from which such station is controlled, and a photocopy of such authorization shall also be posted at all other fixed locations from which the station is controlled. If a photocopy of the authorization is posted at the principal control point, the location of the original shall be stated on that photocopy. In addition, an executed Transmitter Identification Card (FCC Form 452—C) or a plate of metal or other durable substance, legibly indicating the call sign, the licensee's name and address, shall be affixed, readily visible for inspection, to each transmitter operated at a fixed location when such transmitter is not in view of, or is not readily accessible to, the operator of at least one of the locations at which the station authorization or a photocopy thereof is required to be posted.

(b) The current authorization for each station operated as a mobile station shall be retained as a permanent part of the station records, but need not be posted. In addition, an executed Transmitter Identification Card (FCC Form 452—C) or a plate of metal or other durable substance, legibly indicating the call sign and the licensee's name and address, shall be affixed, readily visible for inspection, to each of such transmitters: *Provided*, That, if the transmitter is not in view of the location from which it is controlled, or is not readily accessible for inspection, then such card or plate shall be affixed to the control equipment at the transmitter operating position or posted adjacent thereto.

95.103 Inspection of stations and station records.

All stations and records of stations in the Citizens Radio Service shall be made available for inspection upon the request of an authorized representative of the Commission made to the licensee or to his representative (see 1.6 of this chapter). Unless otherwise stated in this part, all required station records shall be maintained for a period of at least 1 year.

95.105 Current copy of rules required.

Each licensee in this service shall maintain as a part of his station records a current copy of Part 95, Citizens Radio Service, of this chapter.

95.107 Inspection and maintenance of tower marking and lighting, and associated control equipment.

The licensee of any radio station which has an antenna structure required to be painted and illuminated pursuant to the provisions of section 303(q) of the Communications Act of 1934, as amended, and Part 17 of this chapter, shall perform the inspection and maintain the tower marking and lighting, and associated control equipment, in accordance with the requirements set forth in Part 17 of this chapter.

95.111 Recording of tower light inspections.

When a station in this service has an antenna structure which is required to be illuminated, appropriate entries shall be made in the station records in conformity with the requirements set forth in Part 17 of this chapte.

95.113 Answers to notices of violations.

(a) Any licnesee who appears to have violated any provision of the Communications Act or any provision of this chapter shall be served with a written notice calling the facts to his attention and requesting a statement concerning the matter. FCC Form 793 may be used for this purpose.

(b) Within 10 days from receipt of notice or such other period as may be specified, the licensee shall send a written answer, in duplicate, direct to the office of the Commission originating the notice. If an answer cannot be sent nor an acknowledgment made within such period by reason of illness or other unavoidable circumstances, acknowledgment and answer shall be made at the earliest practicable date with a satisfactory e xplanation of the delay.

(c) The answer to each notice shall be complete in itself and shall not be abbreviated by reference to other communications or answers to other notices. In every instance the answer shall contain a statement of the action taken to correct the condition or omission complained of and to preclude its recurrence. If the notice relates to violations that may be due to the physical or electrical characteristics of transmitting apparatus, the licensee must comply with the provisions of 95.53, and the answer to the notice shall state fully what steps, if any, have been taken to prevent future violations, and, if any new apparatus is to be installed, the date such apparatus was ordered, the name of the manufacturer, and the promised date of delivery. If the installation of such apparatus requires a construction permit, the file number of the application shall be given, or if a file number has not been assigned by the Commission, such identification shall be given as will permit ready identification of the application. If the notice of violation relates to lack of attention to or improper operation of the transmitter, the name and license number of the operator in charge, if any, shall also be given.

95.115 False signals.

No person shall transmit false or deceptive communications by radio or identify the station he is operating by means of a call sign which has not been assigned to that station.

95.117 Station location.

(a) The specific location of each Class A base station and each Class A fixed station and the specific area of operation of each Class A mobile station shall be indicated in the application for license. An authorization may be granted for the operation of a Class A base station or fixed station in this service at unspecified temporary fixed locations within a specified general area of operation. However, when any unit or units of a base station or fixed station authorized to be operated at temporary locations actually remains or is intended to remain at the same location for a period of over a year, application for separate authorization specifying the fixed location shall be made as soon as possible but not later than 30 days after the expiration of the 1-year period.

(b) A Class A mobile station authorized in this service may be used or operated anywhere in the United States subject to the provisions of paragraph (d) of this section: *Provided*, That when the area of operation is changed for a period exceeding 7 days, the following procedure shall be observed:

(1) When the change of area of operation occurs inside the same Radio District, the Engineer in Charge of the Radio District involved and the Commission's office, Washington, D.C., 20554, shall be notified.

(2) When the station is moved from one Radio District to another, the Engineers in Charge of

the two Radio Districts involved and the Commission's office, Washington, D.C., 20554, shall be notified.

(c) A Class C or Class D mobile station may be used or operated anywhere in the United States subject to the provisions of paragraph (d) of this section.

(d) A mobile station authorized in this service may be used or operated on any vessel, aircraft, or vehicle of the United States: *Provided,* That when such vessel, aircraft, or vehicle is outside the territorial limits of the United States, the station, its operation, and its operator shall be subject to the governing provisions of any treaty concerning telecommunications to which the United States is a party, and when within the territorial limits of any foreign country, the station shall be subject also to such laws and regulations of that country as may be applicable.

95.119 Control points, dispatch points, and remote control.

(a) A control point is an operating position which is under the control and supervision of the licensee, at which a person immediately responsible for the proper operation of the transmitter is stationed, and at which adequate means are available to aurally monitor all transmissions and to render the transmitter inoperative. Each Class A base or fixed station shall be provided with a control point, the location of which will be specified in the license. The location of the control point must be the same as the transmitting equipment unless the application includes a request for a different location. Exception to the requirement for a control point may be made by the Commission upon specific request and justification therefor in the case of certain unattended Class A stations employing special emissions pursuant to 95.47(e). Authority for such exception must be shown on the license.

(b) A dispatch point is any position from which messages may be transmitted under th supervision of the person at a control point who is responsible for the proper operation of the transmitter. No authorization is required to install dispatch points.

(c) Remote control of a Citizens radio station means the control of the transmitting equipment of that station from any place other than the location of the transmitting equipment, except that direct mechanical control or direct electrical control by wired connections of transmitting equipment from some other point on the same premises, craft, or vehicle shall not be considered remote control. A Class A base or fixed station may be authorized to be used or operated by remote control from another fixed location or from mobile units: *Provided,* That adequate means are available to enable the person using or operating the station to render the transmitting equipment inoperative from each remote control position should improper operation occur.

(d) **Operation of any Class C or Class D station by remote control is prohibited except remote control by wire upon specific authorization by the Commission when satisfactory need is shown.**

[95.119(d) amended eff. 9-15-75; VI (75)-2]

95.121 Civil defense communications.

A licensee of a station authorized under this part may use the licensed radio facilities for the transmission of messages relating to civil defense activities in connection with official tests or drills conducted by, or actual emergencies proclaimed by, the civil defense agency having jurisdiction over the area in which the station is located: *Provided,* That:

(a) The operation of the radio station shall be on a voluntary basis.

(b) [Reserved]

(c) Such communications are conducted under the direction of civil defense authorities.

(d) As soon as possible after the beginning of such use, the licensee shall send notice to the Commission in Washington, D.C., and to the Engineer in Charge of the Radio District in which the station is located, stating the nature of the communications being transmitted and the duration of the special use of the station. In addition, the Engineer in Charge shall be notified as soon as possible of any change in the nature of or termination of such use.

(e) In the event such use is to be a series of pre-planned tests or drills of the same or similar nature which are scheduled in advance for specific times or at certain intervals of time, the licensee may send a single notice to the Commission in Washington, D.C., and to the Engineer in Charge of the Radio District in which the station is located, stating the nature of the communications to be transmitted, the duration of each such test, and the times scheduled for such use. Notice shall likewise be given in the event of any change in the nature of or termination of any such series of tests.

(f) The Commission may, at any time, order the discontinuance of such special use of the authorized facilities.

SUBPART E—
OPERATION OF CITIZENS RADIO STATIONS IN THE UNITED STATES BY CANADIANS

95.131 Basis, purpose and scope.

(a) The rules in this subpart are based on, and are applicable solely to the agreement (TIAS #6931) between the United States and Canada, effective July 24, 1970, which permits Canadian stations in the General Radio Service to be operated in the United States.

(b) The purpose of this subpart is to implement the agreement (TIAS #6931) between the United States and Canada by prescribing rules under which a Canadian licensee in the General Radio Service may operate his station in the United States.

95.133 Permit required.

Each Canadian licensee in the General Radio Service desiring to operate his radio station in the United States, under the provisions of the agreement (TIAS #6931), must obtain a permit for such operation from the Federal Communications Commission. A permit for such operation shall be issued only to a person holding a valid license in the General Radio Service issued by the appropriate Canadian governmental authority.

95.135 Application for permit.

(a) Application for a permit shall be made on FCC Form 410—B. Form 410—B may be obtained from the Commission's Washington, D.C., office or from any of the Commission's field offices. A separate application form shall be filed for each station or transmitter desired to be operated in the United States.

(b) The application form shall be completed in full in English and signed by the applicant. The application must be filed by mail or in person with the Federal Communications Commission, Gettysburg, Pa. 17326, U.S.A. To allow sufficient time for processing, the application should be filed at least 60 days before the date on which the applicant desires to commence operation.

(c) The Commission, at its discretion, may require the Canadian licensee to give evidence of his knowledge of the Commission's applicable rules and regulations. Also the Commission may require the applicant to furnish any additional information it deems necessary.

95.137 Issuance of permit.

(a) The Commission may issue a permit under such conditions, restrictions and terms as it deems appropriate.

(b) Normally, a permit will be issued to expire 1 year after issuance but in no event after the expiration of the license issued to the Canadian licensee by his government.

(c) If a change in any of the terms of a permit is desired, an application for modification of the permit is required. If operation beyond the expiration date of a permit is desired an application for renewal of the permit is required. Application for modification or for renewal of a permit shall be filed on FCC Form 410—B.

(d) The Commission, in its discretion, may deny any application for a permit under this subpart. If an application is denied, the applicant will be notified by letter. The applicant may, within 30 days of the mailing of such letter, request the Commission to reconsider its action.

95.139 Modification or cancellation of permit.

At any time the Commission may, in its discretion, modify or cancel any permit issued under this subpart. In this event, the permittee will be notified of the Commission's action by letter mailed to his mailing address in the United States and the permittee shall comply immediately. A permittee may, within 30 days of the mailing of such letter, request the Commission to reconsider its action. The filing of a request for reconsideration shall not stay the effectiveness of that action, but the Commission may stay its action on its own motion.

95.141 Possession of permit.

The current permit issued by the Commission, or a photocopy thereof, must be in the possession of the operator or attached to the transmitter. The license issued to the Canadian licensee by his government must also be in his possession while he is in the United States.

95.143 Knowledge of rules required.

Each Canadian permittee, operating under this subpart, shall have read and understood this Part 95, Citizens Radio Service.

95.145 Operating conditions.

(a) The Canadian licensee may not under any circumstances begin operation until he has received a permit issued by the Commission.

(b) Operation of station by a Canadian licensee under a permit issued by the Commission must comply with all of the following:

(1) The provisions of this subpart and of Subparts A through D of this part.

(2) Any further conditions specified on the permit issued by the Commission.

95.147 Station identification.

The Canadian licensee authorized to operate his radio station in the United States under the provisions of this subpart shall identify his station by the call sign issued by the appropriate authority of the government of Canada followed by the station's geographical location in the United States as nearly as possible by city and state.

United States of America
Federal Communications Commission

FCC FORM 555—B
April 1976

Temporary Permit
Class D Citizens Radio Station

1
Instructions

- Use this form only if you want a temporary permit while your regular application, FCC Form 505, is being processed by the FCC.
- Do not use this form if you already have a Class D license.
- Do not use this form when renewing your Class D license.

2
Certification
Read, Fill In Blanks, and Sign

I Hereby Certify:

☐ I am at least 18 years of age.

☐ I am not a representative of a foreign government.

☐ I have applied for a Class D Citizens Radio Station License by mailing a completed Form 505 and $4.00 filing fee to the Federal Communications Commission, Gettysburg, PA. 17326

☐ I have not been denied a license or had my license revoked by the FCC.

☐ I am not the subject of any other legal action concerning the operation of a radio station.

_____
Name

_____
Address

_____
Date Form 505 mailed to FCC

_____
Signature

If you cannot certify to the above, you are not eligible for a temporary permit.

Willful false statements void this permit and are punishable by fine and/or imprisonment.

3
Temporary Call Sign

- Complete the blocks as indicated.

Use this temporary call sign until given a call sign by the Federal Communications Commission.

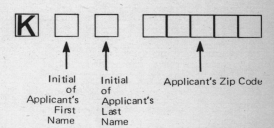

4
Limitations

Your authority under this permit is subject to all applicable laws, treaties and regulations and is subject to the right of use or control by the Government of the United States.

This permit is valid for 60 days from the date the Form 505 is mailed to the FCC.

You must have a temporary permit or a license from the FCC to operate your Citizens Band radio transmitter.

Do Not Mail this form, it is your Temporary Permit.

See the reverse side of this form for a summary of operating instructions.

Using Your Citizens Radio Station

(See Part 95 of FCC Rules & Regulations for complete instructions on authorized station use.)

Welcome to the Citizens Radio Service

Citizens Band Radio is a shared communications service with many people using the same frequencies and channels.

The guidelines provided in this form are not intended as a substitute for FCC Rules, but as a general reference to those operating practices and procedures which will benefit you and other users of Citizens Radio.

Your compliance with these guidelines and your consideration for the rights of others in your radio service is necessary if the full potential and enjoyment of Citizens Radio is to be realized.

1 Who May Operate Your Citizens Radio Station?

You, members of your immediate family living with you, and your employees, while on the job.

2 How Many Transmitters Does this Permit Authorize?

A maximum of five (5).

3 Can the FCC Inspect My Station?

Your station and station records must be available for inspection by an authorized agent of the FCC.

4 Where Should I Keep This Permit?

Keep it in a safe place. Post photocopies at all fixed station locations. Indicate on photocopies the location of this permit. Attach a card with your name, address and temporary call sign to each transmitter.

5 How Shall I Identify My Station?

Identify transmissions in English with your temporary call sign.

6 How Can I Use My Station?

Use it for private short-distance radio-communications for your personal or business activities. Channel 9 is reserved solely for emergency communications and to assist motorists.

Prohibited Communications Include:

Activities contrary to law

Transmitting obscene, indecent or profane messages

Communicating with non-Class D stations

Intentional interference to other radio stations

Transmitting for amusement, entertainment, or over a public address system

Transmitting false distress messages

Advertising, selling, or for hire

7 How High Can My Fixed Station Antenna Be?

See Section 95.37 if your antenna will be over 20 feet above ground. Additional information is available in SS Bulletin 1001-h.

8 May Amplifiers Be Used With My Transmitter?

'Linear' amplifiers are absolutely prohibited. 'Power' microphones may require adjustments to your transmitter.

9 Who Can Make Adjustments to My Transmitters?

Adjustments affecting proper operation may be made only by, or under the supervision of a licensed first or second-class radio operator.

FCC FORM 505
August 1975

United States of America
Federal Communications Commission

Form Approved
GAO No. B-180227(R01 02)

APPLICATION FOR CLASS C OR D STATION LICENSE IN THE CITIZENS RADIO SERVICE

INSTRUCTIONS

A. Print clearly in capital letters or use a typewriter. Put one letter or number per box. Skip a box where a space would normally appear.

B. Enclose appropriate fee with application. Make check or money order payable to Federal Communications Commission. DO NOT SEND CASH. No fee is required of governmental entities. For additional fee details see FCC Form 76-K, or Subpart G of Part 1 of the FCC Rules and Regulations, or you may call any FCC Field Office.

C. Mail application to Federal Communications Commission, Gettysburg, Pa. 17326

NOTICE TO INDIVIDUALS REQUIRED BY PRIVACY ACT OF 1974

Sections 301, 303 and 308 of the Communications Act of 1934 and any amendments thereto (licensing powers) authorize the FCC to request the information on this application. The purpose of the information is to determine your eligibility for a license. The information will be used by FCC staff to evaluate the application, to determine station location, to provide information for enforcement and rulemaking proceedings and to maintain a current inventory of licensees. No license can be granted unless all information requested is provided.

---

1. Complete ONLY if license is for an Individual or Individual Doing Business AS

FIRST NAME   INIT   LAST NAME

2. DATE OF BIRTH
MONTH   DAY   YEAR

3. Complete ONLY if license is for a business, an organization, or Individual Doing Business AS

NAME OF BUSINESS OR ORGANIZATION

4. Mailing Address

4A. NUMBER AND STREET

4B. CITY   **4C. STATE**   **4D. ZIP CODE**

NOTE: Do not operate until you have your own license. Use of any call sign not your own is prohibited

5. If you gave a P.O. Box No., RFD No., or General Delivery in Item 4A, you must also answer items 5A, 5B, and 5C.

5A. NUMBER AND STREET WHERE YOU OR YOUR PRINCIPLE STATION CAN BE FOUND
(If your location can not be described by number and street, give other description, such as, on RT. 2, 3 mi., north of York.)

5B. CITY   **5C. STATE**

6. Type of Applicant (Check Only One Box)

☐ Individual   ☐ Association   ☐ Corporation
☐ Business Partnership   ☐ Governmental Entity
☐ Sole Proprietor or Individual/Doing Business As
☐ Other (Specify) _____

7. This application is for

☐ New License
☐ Renewal
☐ Increase in Number of Transmitters

IMPORTANT
Give Official FCC Call Sign

8. This application is for (Check Only One Box)

☐ Class C Station License
(NON-VOICE—REMOTE CONTROL OF MODELS)
☐ Class D Station License (VOICE)

9. Indicate number of transmitters applicant will operate during the five year license period (Check Only One Box)

☐ 1 to 5   ☐ 6 to 15   ☐ 16 or more (Specify No. and attach statement justifying need.)

10. Certification   I certify that:

• The applicant is not a foreign government or a representative thereof.

• The applicant has or has ordered a current copy of Part 95 of the Commission's rules governing the Citizens Radio Service. See reverse side for ordering information.

• The applicant will operate his transmitter in full compliance with the applicable law and current rules of the FCC and that his station will not be used for any purpose contrary to Federal, State, or local law or with greater power than authorized.

• The applicant waives any claim against the regulatory power of the United States relative to the use of a particular frequency or the use of the medium of transmission of radio waves because of any such previous use, whether licensed or unlicensed.

WILLFUL FALSE STATEMENTS MADE ON THIS FORM OR ATTACHMENTS ARE PUNISHABLE BY FINE AND IMPRISONMENT. U.S. CODE, TITLE 18, SECTION 1001.

11. _____
Signature of: Individual applicant, partner, or authorized person on behalf of a governmental entity, or an officer of a corporation or association

12. Date _____

Sometimes it becomes necessary to return an application. By putting your name and address in the area below. you will enable us to return quickly any application which needs correction or clarification: 1) Put your name on the first line in regular order (for example, Joe Doe); 2) Put your number and street on the second line; 3) Put your city, state, and zip code on the third line.

If necessary, use abbreviations to stay within the guidemarks provided.

appendix c

FCC BUREAU OFFICES

FCC FIELD OFFICES

District No.	Location
1	**BOSTON, MASSACHUSETTS** 1600 Customhouse 165 State Street Boston, MA 02109 Phone: (617) 223-6608 FTS 8-223-6608
2	**NEW YORK, NEW YORK** 201 Varick Street New York, NY 10014 Phone: (212) 620-3437 FTS 8-660-3437
3	**PHILADELPHIA, PENNSYLVANIA** 11425 James A. Byrne Federal Courthouse 601 Market Street Philadelphia, PA 19106 Phone: (215) 597-4411 FTS 8-597-4411
4	**BALTIMORE, MARYLAND** 819 Federal Building 31 Hopkins Plaza Baltimore, MD 21201 Phone: (301) 962-2727 FTS 8-922-2727
5	**NORFOLK, VIRGINIA** Military Circle 870 N. Military Highway Norfolk, VA 23502 Phone: (804) 461-4000 FTS 8-939-6611
6	**ATLANTA, GEORGIA** 1365 Peachtree Street, N.E. Room 440, Massell Building Atlanta, GA 30309 Phone: (404) 526-6381 FTS 8-285-5104
6A	**SAVANNAH, GEORGIA (Sub Office)** Room 238, Federal Building & Courthouse P.O. Box 8004 125 Bull Street Savannah, GA 31402 Phone: (912) 232-4321 FTS 8-287-4320
7	**MIAMI, FLORIDA** 51 S.W. First Avenue Room 919 Miami, FL 33130 Phone: (305) 350-5541 FTS 8-350-5541
7A	**TAMPA, FLORIDA (Sub Office)** 738 Federal Office Bldg. 500 Zack Street Tampa, FL 33602 Phone: (813) 228-2605 FTS 8-826-2606
8	**NEW ORLEANS, LOUISIANA** 829 F. Edward Hebert Federal Building 600 South Street New Orleans, LA 70130 Phone: (504) 589-2094 FTS 8-682-2094
9	**HOUSTON, TEXAS** New Federal Office Bldg. Room 5636 515 Rusk Avenue Houston, TX 77002 Phone: (713) 226-4306 FTS 8-527-4306
9A	**BEAUMONT, TEXAS (Sub Office)** 323 Federal Building 300 Willow Street Beaumont, TX 77701 Phone: (713) 838-0271 FTS 8-527-2317
10	**DALLAS, TEXAS** Earle Cabell Federal Bldg. U.S. Courthouse, Room 13E7 1100 Commerce Street Dallas, TX 75242 Phone: (214) 749-3244 FTS 8-749-3244
11	**LOS ANGELES, CALIFORNIA** 3711 Long Beach Boulevard Room 501 Long Beach, CA 90807 Phone: (213) 426-4451 FTS 8-796-2402
11A	**SAN DIEGO, CALIFORNIA (Sub Office)** Fox Theatre Building 1245 Seventh Avenue San Diego, CA 92101 Phone: (714) 293-5460 FTS 8-895-5460
12	**SAN FRANCISCO, CALIFORNIA** 323-A Customhouse 555 Battery Street San Francisco, CA 94111 Phone: (415) 556-7700 FTS 8-556-7700
13	**PORTLAND, OREGON** 1782 Federal Office Building 1220 S.W. 3rd Avenue Portland, OR 97204 Phone: (503) 221-3097 FTS 8-423-3097
14	**SEATTLE, WASHINGTON** 3256 Federal Building 915 Second Avenue Seattle, WA 98714 Phone: (206) 442-7653 FTS 8-399-7653
15	**DENVER, COLORADO** Suite 2925, The Executive Tower 1405 Curtis Street Denver, CO 80202 Phone: (303) 837-4053 FTS 8-327-4053
16	**ST. PAUL, MINNESOTA** 691 Federal Bldg. & U.S. Courthouse 316 North Robert Street St. Paul, MN 55101 Phone: (612) 725-7819 FTS 8-725-7819
17	**KANSAS CITY, MISSOURI** 1703 Federal Building 601 East 12th Street Kansas City, MO 64106 Phone: (816) 374-5526 FTS 8-758-5526
18	**CHICAGO, ILLINOIS** 230 S. Dearborn St., Rm. 3935 Chicago, IL 60604 Phone: (312) 353-0195 FTS 8-353-0195
19	**DETROIT, MICHIGAN** 1054 Federal Building Washington Blvd. & LaFayette Street Detroit, MI 48226 Phone: (313) 226-6077 FTS 8-226-6879
20	**BUFFALO, NEW YORK** 1307 Federal Building 111 West Huron Street Buffalo, NY 14202 Phone: (716) 842-3216 FTS 8-432-3216
21	**HONOLULU, HAWAII** 502 Federal Building P.O. Box 1021 355 Merchant Street Honolulu, HI 96808 Phone: (808) 546-5640 FTS 8-546-5640
22	**SAN JUAN, PUERTO RICO** 322-323 Federal Building SPO Box 2987 Comercio, San Justo & Tanca Streets San Juan, Puerto Rico Phone: (809) 722-4562 No FTS at present
23	**ANCHORAGE, ALASKA** U.S.P.O. & Courthouse Bldg. Room G-63 4th & F Streets Anchorage, AK 99510 Phone: (907) 265-5201 FTS 8-265-5201
24	**WASHINGTON, D.C.** 1919 M St., N.W. Rm. 411 Washington, DC 20554 Phone: (202) 632-7000 FTS 8-632-7000

appendix D

FCC MONITORING STATIONS

ALLEGAN, MICHIGAN
Allegan Monitoring Station
P.O. Box 89
Allegan, MI 49010
Phone: (616) 673-2063

ANCHORAGE, ALASKA
Anchorage Monitoring Station
P.O. Box 6303
Anchorage, AK 99502
Phone: (907) 344-1011

BELFAST, MAINE
Belfast Monitoring Station
P.O. Box 470
Belfast, ME 04915
Phone: (207) 338-4088

CANANDAIGUA, NEW YORK
Canandaigua Monitoring Station
P.O. Box 374
Canandaigua, NY 14424
Phone: (315) 394-4240

CHILLICOTHE, OHIO
Chillicothe Monitoring Station
P.O. Box 251
Chillicothe, OH 45601
Phone: (614) 775-6523

DOUGLAS, ARIZONA
Douglas Monitoring Station
P.O. Box 6
Douglas, AZ 85607
Phone: (602) 364-2133

FERNDALE, WASHINGTON
Ferndale Monitoring Station
P.O. Box 1125
Ferndale, WA 98248
Phone: (206) 354-4892

FORT LAUDERDALE, FLORIDA
Fort Lauderdale Monitoring Station
9900 West State Road 84
P.O. Box 22836
Fort Lauderdale, FL 33316
Phone: (305) 472-5511
FTS 8-350-9352

LAUREL, MARYLAND
Laurel Monitoring Station
7435 Oakland Mills Road
P.O. Box B
Laurel, MD 20810
Phone: (301) 725-3474

LIVERMORE, CALIFORNIA
Livermore Monitoring Station
P.O. Box 311
Livermore, CA 94550
Phone: (415) 447-3614

POWDER SPRINGS, GEORGIA
Powder Springs Monitoring Station
3600 Hiram-Lithia Spring Road, SW
P.O. Box 85
Powder Springs, GA 30073
Phone: (404) 943-5420

SABANA SECA, PUERTO RICO
Sabana Seca Monitoring Station
P.O. Box 181
Sabana Seca, Puerto Rico 00749
Phone: (809) 784-3772

WAIPAHU, HAWAII
Waipahu Monitoring Station
P.O. Box 1035
Waipahu, HI 96797
Phone: (808) 677-3954

GRAND ISLAND, NEBRASKA
Grand Island Monitoring Station
P.O. Box 1588
Grand Island, NB 68801
Phone: (308) 382-4296
FTS 8-864-2289

KINGSVILLE, TEXAS
Kingsville Monitoring Station
P.O. Box 632
Kingsville, TX 78363
Phone: (512) 592-2531